ENERGY SCIENCE, ENGINEERING AND TECHNOLOGY

AGRICULTURE-BASED BIOFUELS AND THE FEASIBILITY OF BIOMASS-BASED DIESEL AND JET FUEL

ENERGY SCIENCE, ENGINEERING AND TECHNOLOGY

Additional books in this series can be found on Nova's website under the Series tab.

Additional e-books in this series can be found on Nova's website under the e-book tab.

ENERGY SCIENCE, ENGINEERING AND TECHNOLOGY

AGRICULTURE-BASED BIOFUELS AND THE FEASIBILITY OF BIOMASS-BASED DIESEL AND JET FUEL

DOUG R. MONTY
EDITOR

New York

Library of Congress Cataloging-in-Publication Data

ISBN: 978-1-63117-783-5

Published by Nova Science Publishers, Inc. † New York

CONTENTS

PREFACE

Increasing dependence on foreign sources of crude oil, concerns over global climate change, and the desire to promote domestic rural economies have raised interest in renewable biofuels as an alternative to petroleum in the U.S. transportation sector. However, energy from renewable sources has historically been more expensive to produce and use than fossil-fuel-based energy. This book describes agriculture-based biofuels and the evolution of the U.S. biofuels sector with a focus on the role that federal policy has played in shaping its development. It highlights emerging issues that are critical to the biofuels sector and of relevance to Congress. This book also summarizes the best available public data on the production, capacity, cost, market demand, and feedstock availability for the production of biomass-based diesel and jet fuel. It includes an overview of the current conversion processes and current state-of-development for the production of biomass-based jet and diesel fuel, as well as the key companies pursuing this effort.

Chapter 1 - Since the late 1970s, U.S. policymakers at both the federal and state levels have authorized a variety of incentives, regulations, and programs to encourage the production and use of agriculture-based biofuels—i.e., any fuel produced from biological materials. Initially, federal biofuels policies were developed to help kick-start the biofuels industry during its early development, when neither production capacity nor a market for the finished product was widely available. Federal policy (e.g., tax credits, import tariffs, grants, loans, and loan guarantees) has played a key role in helping to close the price gap between biofuels and cheaper petroleum fuels. Now, as the industry has evolved, other policy goals (e.g., national energy security, climate change concerns, support for rural economies) are cited by proponents as justification for continuing or enhancing federal policy support.

The U.S. biofuels sector responded to these government incentives by expanding output every year from 1980 through 2011 (with the exception of 1996), with important implications for the domestic and international food and fuel sectors. Production of the primary U.S. biofuel, ethanol (derived from corn starch), has risen from about 175 million gallons in 1980 to nearly 14 billion gallons in 2011. U.S. biodiesel production (derived primarily from vegetable oil), albeit much smaller, has also shown strong growth, rising from 0.5 million gallons in 1999 to a record 969 million gallons in 2012. Despite the rapid growth of the past decades, total agriculture-based biofuels consumption accounted for only about 8% of U.S. transportation fuel consumption (9.7% of gasoline and 1.5% of diesel) in 2012.

Federal biofuels policies have had costs, including unintended market and environmental consequences and large federal outlays (estimated at $7.7 billion in 2011, but declining to $1.3 billion in 2012 with the expiration of the ethanol blender's tax credit). Despite the direct and indirect costs of federal biofuels policy and the relatively small role of biofuels as an energy source, the U.S. biofuels sector continues to push for federal involvement. But critics of federal policy intervention in the biofuels sector have also emerged. Current issues and policy developments related to the U.S. biofuels sector that are of interest to Congress include

- Many federal biofuels policies require routine congressional monitoring and occasional reconsideration in the form of reauthorization or new appropriations.
- The 10% ethanol-to-gasoline blend ratio—known as the "blend wall"—poses a barrier to expansion of ethanol use. The Environmental Protection Agency (EPA) issued waivers to allow ethanol blending of up to 15% (per gallon of gasoline) for use in model year 2001 and newer light-duty motor vehicles. However, the limitation to newer vehicles, coupled with infrastructure issues, could limit rapid expansion of blending rates.
- The slow development of cellulosic biofuels has raised concerns about the industry's ability to meet large federal usage mandates, which in turn has raised the potential for future EPA waivers of mandated biofuel volumes and has contributed to a cycle of slow investment in and development of the sector.

In 2012, the expiration of the blender tax credit, poor profit margins (due primarily to high corn prices), and the emerging blend wall limitation have contributed to a drop-off in ethanol production and have generated considerable uncertainty about the ethanol industry's future.

Chapter 2 - This study summarizes the best available public data on the production, capacity, cost, market demand, and feedstock availability for the production of biomass-based diesel and jet fuel. It includes an overview of the current conversion processes and current state-of-development for the production of biomass-based jet and diesel fuel, as well as the key companies pursuing this effort. The discussion analyzes all this information in the context of meeting the Renewable Fuels Standard mandate, highlights uncertainties for the future industry development, and key business opportunities.

In: Agriculture-Based Biofuels ...
Editor: Doug R. Monty

ISBN: 978-1-63117-783-5

Chapter 1

AGRICULTURE-BASED BIOFUELS: OVERVIEW AND EMERGING ISSUES*

Randy Schnepf

SUMMARY

Since the late 1970s, U.S. policymakers at both the federal and state levels have authorized a variety of incentives, regulations, and programs to encourage the production and use of agriculture-based biofuels—i.e., any fuel produced from biological materials. Initially, federal biofuels policies were developed to help kick-start the biofuels industry during its early development, when neither production capacity nor a market for the finished product was widely available. Federal policy (e.g., tax credits, import tariffs, grants, loans, and loan guarantees) has played a key role in helping to close the price gap between biofuels and cheaper petroleum fuels. Now, as the industry has evolved, other policy goals (e.g., national energy security, climate change concerns, support for rural economies) are cited by proponents as justification for continuing or enhancing federal policy support.

The U.S. biofuels sector responded to these government incentives by expanding output every year from 1980 through 2011 (with the exception of 1996), with important implications for the domestic and international food and fuel sectors. Production of the primary U.S. biofuel, ethanol (derived from

* This is an edited, reformatted and augmented version of the Congressional Research Service Publication, CRS Report for Congress R41282, dated May 1, 2013.

corn starch), has risen from about 175 million gallons in 1980 to nearly 14 billion gallons in 2011. U.S. biodiesel production (derived primarily from vegetable oil), albeit much smaller, has also shown strong growth, rising from 0.5 million gallons in 1999 to a record 969 million gallons in 2012. Despite the rapid growth of the past decades, total agriculture-based biofuels consumption accounted for only about 8% of U.S. transportation fuel consumption (9.7% of gasoline and 1.5% of diesel) in 2012.

Federal biofuels policies have had costs, including unintended market and environmental consequences and large federal outlays (estimated at $7.7 billion in 2011, but declining to $1.3 billion in 2012 with the expiration of the ethanol blender's tax credit). Despite the direct and indirect costs of federal biofuels policy and the relatively small role of biofuels as an energy source, the U.S. biofuels sector continues to push for federal involvement. But critics of federal policy intervention in the biofuels sector have also emerged. Current issues and policy developments related to the U.S. biofuels sector that are of interest to Congress include

- Many federal biofuels policies require routine congressional monitoring and occasional reconsideration in the form of reauthorization or new appropriations.
- The 10% ethanol-to-gasoline blend ratio—known as the "blend wall"—poses a barrier to expansion of ethanol use. The Environmental Protection Agency (EPA) issued waivers to allow ethanol blending of up to 15% (per gallon of gasoline) for use in model year 2001 and newer light-duty motor vehicles. However, the limitation to newer vehicles, coupled with infrastructure issues, could limit rapid expansion of blending rates.
- The slow development of cellulosic biofuels has raised concerns about the industry's ability to meet large federal usage mandates, which in turn has raised the potential for future EPA waivers of mandated biofuel volumes and has contributed to a cycle of slow investment in and development of the sector.

In 2012, the expiration of the blender tax credit, poor profit margins (due primarily to high corn prices), and the emerging blend wall limitation have contributed to a drop-off in ethanol production and have generated considerable uncertainty about the ethanol industry's future.

INTRODUCTION

Increasing dependence on foreign sources of crude oil, concerns over global climate change, and the desire to promote domestic rural economies have raised interest in renewable biofuels as an alternative to petroleum in the U.S. transportation sector. However, energy from renewable sources has historically been more expensive to produce and use than fossil-fuel-based energy.[1] U.S. policymakers have attempted to overcome this economic impediment by enacting an increasing number of policies since the late 1970s, at both the state and federal levels, to directly support U.S. biofuels production and use. Policy measures have included blending and production tax credits to lower the cost of biofuels to end users, an import tariff to protect domestic ethanol from cheaper foreign-produced ethanol, research grants to stimulate the development of new technologies, loans and loan guarantees to facilitate the development of biofuels production and distribution infrastructure, and, perhaps most importantly, minimum usage requirements to guarantee a market for biofuels irrespective of their cost.[2]

This report describes agriculture-based biofuels and the evolution of the U.S. biofuels sector with a focus on the role that federal policy has played in shaping its development.[3] In addition, it highlights emerging issues that are critical to the biofuels sector and of relevance to Congress.

BIOFUELS DEFINED

Any fuel produced from biological materials—whether burned for heat or processed into alcohol—qualifies as a "biofuel." The term is most often used to refer to liquid transportation fuels produced from some type of biomass. The two principal biofuels are ethanol and biodiesel; however, other fuels such as methanol and butanol could also qualify when produced from a qualifying biomass.

Biomass is organic matter that can be converted into energy. Common examples of biomass include food crops, energy crops (e.g., switchgrass or prairie perennials), crop residues, wood waste and byproducts, and animal manure. The term biomass has been a part of legislation enacted by Congress for various programs over the past 30 years; however, its explicit definition has evolved with shifting policy objectives.[4] Over the last few years, the concept of biomass has grown to include such diverse sources as algae, construction debris, municipal solid waste, yard waste, and food waste. The exact definition of biomass is critical, since it determines which feedstocks and resultant biofuels qualify for the different federal biofuels programs.

For example, the principal biofuels program in effect as of this report is the Renewable Fuels Standard (RFS), which mandates annual usage rates for four nested categories of biofuels—(1) total renewable fuels, (2) advanced renewable fuels, (3) cellulosic biofuel, and (4) biomass-based diesel.[5] Qualifying biofuels under each category are differentiated by their type of feedstock, the land on which the feedstock is produced (e.g., federal versus private, virgin versus previously cultivated soil, etc.), the production process used both to grow the feedstock and to process it into a biofuel (certain technologies are favored based primarily on environmental considerations), and the estimated amount of greenhouse gas emissions that result from the entire production pathway.

The idea of formally defining biomass has evoked criticism. Some argue that by explicitly enunciating qualifying feedstocks, the definition may be excluding new or as-yet-undiscovered feedstocks that may emerge in the future. Also, there appears to be some inconsistency across programs. For example, algae-based biofuels presently do not qualify for inclusion under the RFS cellulosic biofuels mandate, but do qualify for the "advanced other" biofuels mandate, as well as for the cellulosic biofuels tax credit and the depreciation allowance for qualifying cellulosic biofuels plants.[6] These differentiations tend to confuse and may slow or inhibit investments in algae-based biofuels.

Ethanol from Corn Starch Dominates U.S. Biofuels Production

Ethanol is the principal biofuel produced in the United States (Figure 1). Ethanol, or ethyl alcohol, is an alcohol made by fermenting and distilling simple sugars. As a result, ethanol can be produced from any biological feedstock that contains appreciable amounts of sugar or materials that can be converted into sugar such as starch or cellulose. Sugar beets and sugar cane are examples of feedstock that contain sugar. Corn contains starch that can relatively easily be converted into sugar. Trees, grasses, and most agricultural and municipal wastes are made up of a significant percentage of cellulose, which can also be converted to sugar, although with more difficulty than is required to convert starch.

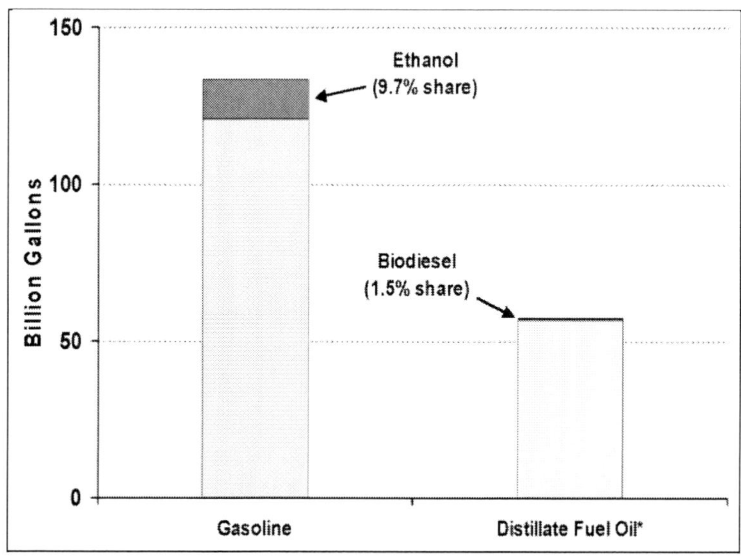

Sources: Calculated by CRS based on data from the Energy Information Agency (EIA), Department of Energy (DOE), *Monthly Energy Review*, March 2013: ethanol from Table 10.3, biodiesel from Table 10.4, and gasoline and distillate fuel oil use from Table 3.5.

Note: All data are in actual volumes; i.e., there is no conversion for gasoline energy equivalency. Distillate fuel oil includes both transportation and home heating oil uses.

Figure 1. Ethanol Had Nearly a 10% Share of U.S. Motor Gasoline Fuel Use in 2012.

Since its development in the late 1970s, U.S. biofuels output has relied almost exclusively on ethanol produced from corn starch. Small amounts of ethanol have also been produced using sorghum, wheat, barley, and brewery waste. This contrasts with Brazil, the world's second-largest ethanol producer behind the United States, where sugar cane is the principal feedstock. In 2012, the United States and Brazil accounted for 88% of the world's ethanol production.[7] Approximately 13.3 billion gallons of ethanol were produced in the United States in 2012, over 95% from corn starch.

Because of concerns over the significant expansion in corn production for use as an ethanol feedstock, interest has grown in spurring the development of motor fuels produced from cellulosic biomass materials. Since these biomass sources do not compete with traditional food and feed crops for prime cropland, it is thought that their use would result in substantially fewer unintended market effects. However, the technology needed for the conversion of cellulose into its constituent sugars before conversion to biofuels, while successful in laboratory settings, is thought to be expensive relative to corn ethanol and has yet to be replicated on a significant commercial scale.[8] Many uncertainties remain concerning both the viability and the speed of commercial development of cellulosic biofuels.[9]

After ethanol, biodiesel is the next most significant biofuel in the United States. Biodiesel is an alternative diesel fuel that can be produced from any type of organic-based oil, including vegetable oils, animal fats, and waste restaurant grease and oils. In the United States and Brazil, biodiesel has traditionally been made from soybean oil. In the European Union, rapeseed oil is the primary feedstock, while Canada relies primarily on canola oil. In recent years persistently high vegetable oil prices have pushed biodiesel producers to increase the share of much cheaper animal fats (especially poultry fat) and tropical palm oil; however, soybean oil remains the largest single source of biodiesel feedstock in the United States, with a share of over 56% in 2012.[10]

Other biofuels with the potential to play a role in the U.S. market include diesel fuel substitutes and other alcohols (e.g., methanol and butanol) produced from biomass.

Biofuels Value Determinants

The value of a biofuel is determined by its end use. Ethanol is primarily used as a substitute for gasoline; however, it has some additional properties (i.e., as an oxygenate and an octane enhancer) that provide value as a gasoline additive. Biodiesel's primary use is as a substitute for petroleum-based diesel transportation fuel; however, biodiesel can also be used as a direct substitute for home heating oil and as a blend in jet fuel. Also, both ethanol and biodiesel may derive additional value as an additive to meet federal usage mandates under the Renewable Fuel Standard (RFS) depending on market conditions.

The Renewable Fuel Standard (RFS)[11]

The RFS requires the blending of renewable fuels (including ethanol and biodiesel) in U.S. transportation fuel. The RFS includes specific quotas for total renewable biofuels, as well as nested subcategories for advanced biofuels (i.e., non-corn-starch ethanol), cellulosic biofuels, and biomass-based diesel fuel. The RFS also includes a cap on the eligible volume of corn-starch ethanol.[12] The RFS is administered by EPA. Qualifying biofuels must meet explicit criteria on lifecycle greenhouse gas (GHG) emissions[13] and feedstock production pathways (including restrictions on the land on which feedstocks are produced, feedstock production methods, and the biofuels plant processing technology).

Federal policy that mandates the use of a minimum volume of biofuel creates a source of demand that is not based on price, but rather on government fiat. As long as the consumption of biofuels is less than the mandated volume, its use is obligatory.

Ethanol Sources of Demand

With respect to ethanol, there is no difference to the end user between corn-starch ethanol, sugarcane ethanol, and cellulosic ethanol, although their production processes differ substantially in terms of feedstock, technology, and cost. As a result, all three share the same value determinants. In the presence of government policy, demand for ethanol derives from four potential uses:

- as an oxygenate additive in gasoline to help improve engine combustion and cleaner burning of fuel;
- as an additive to gasoline to enhance its octane level and engine performance;[14]
- as an additive to gasoline at blend ratios of up to 10% ethanol and 90% gasoline (known as E10), to meet federally mandated minimum usage requirements under one of the RFS categories for qualifying ethanol biofuels;[15] or
- as a substitute for gasoline at ethanol-to-gasoline blend ratios greater than E10.

Biodiesel Sources of Demand

In the presence of government policy, demand for biodiesel derives from the following potential uses:

- as a substitute for petroleum-based diesel transportation fuel;
- as a substitute for home heating oil;
- as a blend in jet fuel; and
- as an additive to petroleum-based diesel to meet federally mandated minimum usage requirements under one of the RFS categories for qualifying biofuels.[16]

Biofuel Supply Relative to RFS Mandates Affects Valuation

Depending on the relationship between the RFS mandate (blending demand) and the available supply (production plus imports) of qualifying biofuels, different RFS biofuels categories may have significantly different valuations, as greater scarcity will lead to greater value.

Under the RFS, each gallon of qualifying biofuel has an associated renewable identification number (RIN) that is detached at point of blending and submitted to the EPA as proof of fulfilling that year's RFS usage requirement for a specific biofuel category.[17] When a specific biofuel is blended (or used) in excess of its RFS mandate, the surplus RINs may be sold (ideally to another fuel blender to make up for a shortfall in meeting that blender's own RFS mandate) or stored for use in meeting the following year's RFS mandate. As a result of their tradability, secondary markets for RINs—by RFS category—have developed and gain in importance whenever the supply of a specific biofuel type tightens relative to its RFS mandate. RIN values are nested— since cellulosic and biomass-based diesel RINs can be used to meet their own category as well as the advanced and total categories, they have an inherent premium over advanced and total RINs. Similarly, advanced RINs would have a premium over total RINs.

In contrast, when the supply of a specific biofuels category exceeds its mandated usage volume, the associated "nested" value will diminish. In volumes above the RFS total renewable mandate, biofuels use is no longer obligatory and it must compete directly in the marketplace with its petroleum-based counterpart. As a result, once they have met their RFS blending mandates, fuel blenders, seeking to maximize their profits, are very sensitive to price relationships between petroleum-based fuels and biofuels. This is particularly important for ethanol since it contains only about 68% of the energy content of gasoline. As a result, value-conscious consumers could be expected to willingly pay only about 68% of the price of gasoline for ethanol.

From 2006—when the RFS was first introduced—through 2011, both ethanol production capacity, supply (production and imports combined), and consumption have easily exceeded the federally mandated usage levels (Figure 2).[18] As a result, ethanol's marginal value during that period was as a transportation fuel (rather than as an additive), where it competed directly with gasoline. However, economic conditions changed substantially in 2012, driven largely by the severe drought that summer, and the RFS has played a larger role in driving ethanol use. As for biodiesel, which is significantly more expensive to produce than its petroleum-based counterpart, biodiesel's use has been driven almost entirely by federal policy—i.e., the RFS biomass-based diesel and the biodiesel production tax credit (described below).

Blend Wall Emerges As Major Value Determinant

An important valuation concern for U.S. ethanol consumption in 2013 is the emergence of the so-called "blend wall" as a constraint on domestic consumption of ethanol in sufficient volumes to satisfy the RFS mandate. Ethanol-gasoline blends of up to 10% ethanol are compatible with existing vehicles and infrastructure (fuel tanks, retail pumps, delivery infrastructure, etc.). All automakers that produce cars and light trucks for the U.S. market warranty their vehicles to run on gasoline with up to 10% ethanol (E10); however, automakers have been reluctant to offer such warranties for higher ethanol blend ratios. As a result, the 10% blend ratio represents an upper bound (sometimes referred to as the "blend wall") to the amount of ethanol that can be introduced into the gasoline pool given the current automobile fleet and fuel delivery infrastructure.

In 2012, ethanol accounted for nearly a 10% share of blended gasoline sold in the United States (Figure 1). In 2013, the RFS mandates for non-advanced ethanol of 13.8 bgals will likely exceed the blend wall (estimated at approximately 13 bgals by CRS based on EIA data). Supplementing actual ethanol blending with carry-over RINs (estimated at 2.6 bgals) will likely be sufficient to satisfy the 2013 RFS; however, surmounting the blend wall could prove more difficult in 2014.[19] Because of this infrastructure constraint, ethanol production in excess of the blend wall will have limited value in the domestic market unless it is consumed at higher blending ratios in flex-fuel vehicles (FFVs) or exported into the international market.[20]

EVOLUTION OF THE U.S. ETHANOL SECTOR

Federal Policy Kick-Starts Ethanol Production

Several events contributed to the startup and growth of U.S. ethanol production in the late 1970s. First, the global energy crises of the early and late 1970s provided the rationale for a federal policy initiative aimed at promoting energy independence from foreign crude oil sources. In response, the U.S. Congress established a partial exemption for ethanol from the motor fuels excise tax (legislated as part of the Energy Tax Act of 1978). All ethanol blended in the United States—whether imported or produced domestically—

was eligible for a $0.40 per gallon tax credit. In 1980, an import duty for fuel ethanol was established by the Omnibus Reconciliation Act of 1980 (P.L. 96-499) to offset the domestic tax credit being applied to foreign-sourced ethanol.

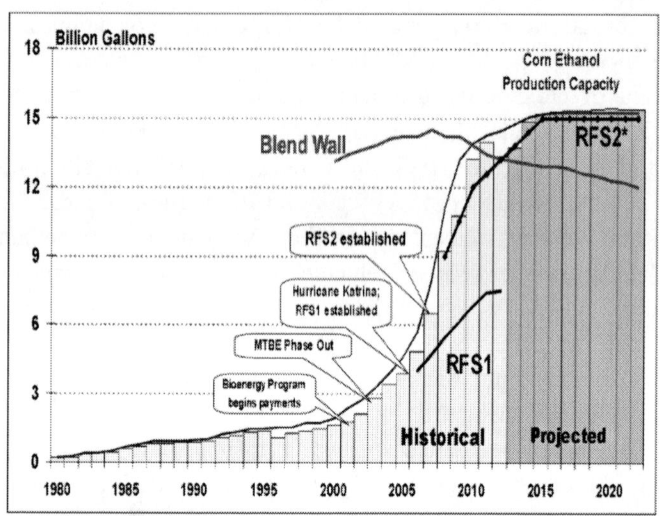

Source: Ethanol consumption historical data for 1980-2012 is from EIA, *Monthly Energy Review*, March 2013, Table 3.5; blend wall historical and projected data are calculated by CRS based on the EIA, DOE, data found in the AEO *Annual Energy Outlook* 2013. Projections for 2013-2022 are corn ethanol production from FAPRI, *FAPRI- MU Biofuel Baseline*, FAPRI-MU Report #02-13, March 2013. The projection data exclude ethanol production from advanced sources, e.g., cellulosic or sugarcane.

Note: RFS2* shown in the chart represents the non-advanced component (RFS code D6) equal to the total renewable fuel mandate minus the advanced biofuel mandate, and roughly approximates the cap on qualifying corn-starch ethanol consumption; ethanol from advanced sources are excluded from this data and this chart. Achieving the corn ethanol consumption levels in excess of the blend wall (as portrayed in this chart and described later in the text) would necessitate substantial consumption at higher blends such as E15 or E85.

Figure 2. U.S. Corn Ethanol Consumption, RFS, and Blend Wall, 1980 to 2022.

As U.S. ethanol production began to emerge in the 1980s, ethanol became recognized as a gasoline oxygenate. The Deficit Reduction Act of 1984 raised the ethanol tax credit to $0.60 per gallon.[21] Based on its oxygenate characteristic, provisions of the Clean Air Act Amendments of 1990 (CAAA90) favored ethanol blending with reformulated gasoline (RFG).[22] One of the requirements of RFG specified by CAAA90 was a 2% oxygen requirement, which was met by blending "oxygenates," including methyl tertiary butyl ether (MTBE) and ethanol into the gasoline.[23] Ethanol was the preferred oxygenate in the Midwest where it was produced, while MTBE—a petroleum derivative—was used in almost all RFG outside of the Midwest.

In addition to CAAA90 oxygenate requirements, a tax credit for small ethanol producer was established in 1990 (Omnibus Budget Reconciliation Act of 1990; P.L. 101-508) as a $0.10 per gallon supplement to the existing ethanol tax credit, but limited to the first 15 million gallons of ethanol produced by ethanol producers with production capacity below 30 million gallons per year.[24] Aided by these events, the U.S. ethanol industry steadily grew during its first two decades—rising from an estimated 175 million gallons in 1980 to 1.8 billion gallons in 2001, when ethanol production was using about 7% of the U.S. corn crop.

Government Role Has Grown Since 2000

The first decade of the 2000s experienced a substantial increase in federal involvement in the U.S. biofuels sector. In FY2001, the Bioenergy Program[25] began making payments from the U.S. Department of Agriculture's (USDA's) Commodity Credit Corporation (CCC)[26] to eligible biofuel producers—ethanol and biodiesel—based on any year-to-year increases in the quantity of biofuels produced. The Bioenergy Program was instituted by USDA because the program's principal goal was to encourage greater purchases of eligible farm commodities used in the production of biofuels (e.g., corn for ethanol or soybean oil for biodiesel).

The executive order creating the Bioenergy Program was followed by a series of legislation containing various provisions that further aided the U.S. biofuels industry. The first of these new laws—the Biomass Research and Development Act of 2000 (Biomass Act; Title III, P.L. 106- 224)—contained several provisions to expand research and development in the area of biomass-based renewable fuel production.

The 2002 farm bill (P.L. 107-171) included several biofuels programs spread across three separate titles—Title II: Conservation, Title VI: Rural Development, and Title IX: Energy (the first-ever energy title in a farm bill). Each title contained programs that encouraged the research, production, and use of renewable fuels such as ethanol, biodiesel, anaerobic digesters, and wind energy systems. In addition, Section 9010 of Title IX codified and extended the Bioenergy Program and its funding by providing that $150 million would be available annually through the CCC for FY2003-FY2006.[27]

The Healthy Forests Restoration Act of 2003 (P.L. 108-148) amended the Biomass Act of 2000 by expanding the use of grants, contracts, and assistance for biomass to include a broader range of forest management activities. It also expanded funding availability of programs established by the Biomass Act and the 2002 farm bill, and it established a program to accelerate adoption of biomass-related technologies through community-based marketing and demonstration activities, and to establish small-scale businesses to use biomass materials.

The American Jobs Creation Act of 2004 (P.L. 108-357) contained a provision (Section 301) that replaced the existing tax exemptions for alcohol fuels (i.e., ethanol) with an excise tax credit of $0.51 per gallon. This act also extended the small ethanol producer tax credit.

MTBE Phase-out Enhances Ethanol's Value

In addition to a growing list of federal and state policies, the U.S. biofuels industry received an additional boost in the early 2000s with the emergence of water contamination problems associated with underground MTBE storage tanks in several locations scattered throughout the country. MTBE was thought to be a possible carcinogen and, as a result, posed serious health and liability issues. In 1999, California (which, at the time, consumed nearly 32% of the MTBE used in the United States) petitioned the U.S. Environmental Protection Agency (EPA) for a waiver of the CAAA90 oxygenate requirement.[28] However, California's waiver request was denied by the EPA in mid-2001 since the EPA determined that there was sufficient ethanol production available to replace MTBE.

By 2003, legislation that would phase out or restrict the use of MTBE in gasoline had been passed in 16 states, including California and New York (with a combined 40% national MTBE market share).[29] Between October 1, 2003, and January 1, 2004, over 43% of MTBE consumption in the United States was banned. According to the EIA, the state MTBE ban would require an additional demand for ethanol of 2.73 billion gallons in 2004.

With the legislative boosts and the MTBE phase-out, investments in the biofuels sector began to show results. The number of plants producing ethanol grew from 50 on January 1, 1999, to 81 by January 1, 2005. Concomitantly, U.S. ethanol production began to accelerate, rising to 3.9 billion gallons by 2005 and using over 14% of the nation's corn crop (Table 1), up from 1.8 bgals and 7% of the corn crop in 2001.

Table 1. U.S. Corn-Use Share of Annual Production by Major Activity, 1980 to 2012

Period	Ethanol	Food	Exports	Feed
1980-1984	2%	11%	30%	64%
1985-1989	4%	14%	26%	46%
1990-1994	5%	14%	21%	58%
1995-1999	5%	14%	21%	55%
2000-2004	10%	14%	18%	60%
2005-2009	25%	11%	18%	55%
2010-2012	41%	12%	12%	37%

Source: Period averages are calculated by CRS from the USDA, PSD database, March 8, 2013.

Note: Values may sum to greater than 100% because some usage may derive from carryover stocks. The table data for the "Feed" and "Export" categories have not been adjusted to include distillers dried grains and solubles (DDGS)—a protein-rich animal feed that is a by-product of corn-based ethanol production.

The Ethanol Industry's Perfect Storm in 2005

On the heels of the large MTBE phase-out that occurred in 2004 and the surge in ethanol demand, two major events coincided in 2005 to produce extremely favorable economic conditions in the U.S. ethanol sector that persisted through most of 2006. These events included the following.

- The Energy Policy Act of 2005 (EPACT; P.L. 109-58) was signed into law on August 8, 2005. EPACT contained several provisions related to agriculture-based renewable energy production, including biofuels research and funding, expansions of existing biofuels tax credits and creation of new credits, and the creation of the first-ever national minimum-usage mandate, the Renewable Fuels Standard (RFS1; Section 1501), which required that 4 billion gallons (bgals) of ethanol be used domestically in 2006, increasing to 7.5 bgals by 2012.
- In August and September 2005, Hurricanes Katrina and Rita struck the Gulf Coast region causing severe damage to local petroleum importing and refining infrastructure, putting them off-line for several months, and driving gasoline prices sharply higher. Meanwhile, corn prices remained relatively low at about $2 per bushel, creating a period of extreme profitability for the ethanol sector.

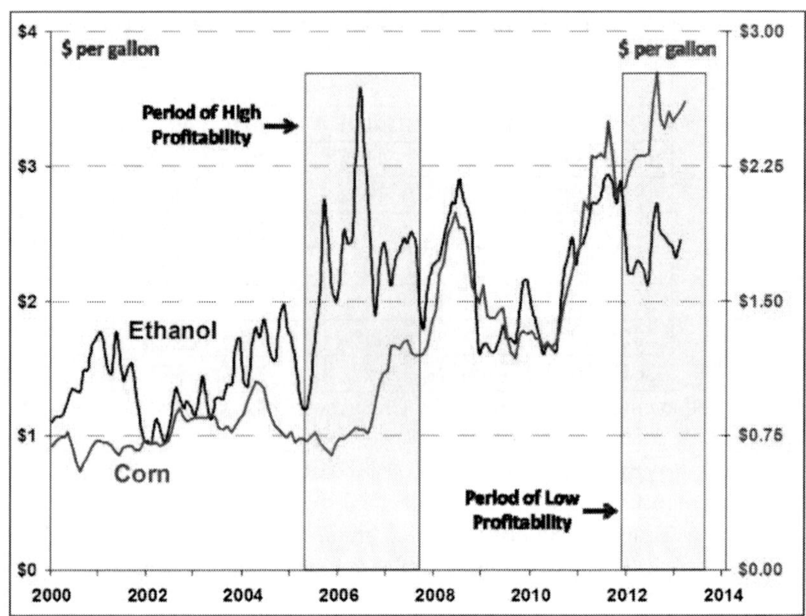

Source: Corn price data are monthly average farm prices, National Agricultural Statistics Service (NASS), USDA; ethanol price is the rack price, f.o.b., Nebraska Ethanol Board, Nebraska Energy Office, Lincoln, NE.

Note: Corn prices ($ per bushel) have been converted to $ per gallon by CRS—i.e., the price of corn used per gallon of ethanol—by dividing the per bushel price by 2.75 (an estimate of gallons of ethanol per bushel of corn).

Figure 3. Comparison of Monthly Prices: Ethanol versus Corn.

The combination of high ethanol prices and relatively low corn prices that began in late 2005 and persisted through 2006 and into 2007 created a period of "unique" profitability for the U.S. ethanol industry (Figure 3). At that time, a 40 million gallon nameplate ethanol plant costing approximately $60 million could recover its entire capital investment in less than a year of normal operations.[30] In addition, the establishment of the first RFS—by guaranteeing a market for new ethanol production—removed much of the investment risk from the sector.

As a result of this "perfect storm" of policy and market events, investment money flowed into the construction of new ethanol plants, and U.S. ethanol production capacity (either in existence or under construction) more than doubled in just four years, rising from an estimated 4.4 bgals produced in 81 plants in January 2005 to 10.6 bgals produced in 170 plants by January 2009. The ethanol expansion was almost entirely in dry-mill corn processing plants.

As a result, corn's role as the primary feedstock used in ethanol production in the United States continued to grow. In 2006, corn use for ethanol nearly matched U.S. corn exports at about 2.1 billion bushels. In 2007, U.S. corn exports hit a record 2.4 billion bushels; however, by then corn-for-ethanol use had jumped to over 3 billion bushels. For the first time in U.S. history, the bushels of corn used for ethanol production would be greater than the bushels of corn exported (Table 1 and Figure 4).

EISA Greatly Expands Mandate, Shifts Focus to Cellulosic Biofuels

In light of the rapid expansion of the U.S. biofuels industry, the RFS1 mandate was outgrown in 2006—the same year it was first implemented (Figure 2). On December 19, 2007, Congress dramatically raised the "bar" by passing the Energy Independence and Security Act of 2007 (EISA, P.L. 110-140).[31] EISA superseded and greatly expanded EPACT's biofuels mandate relative to historical production (Figure 5).

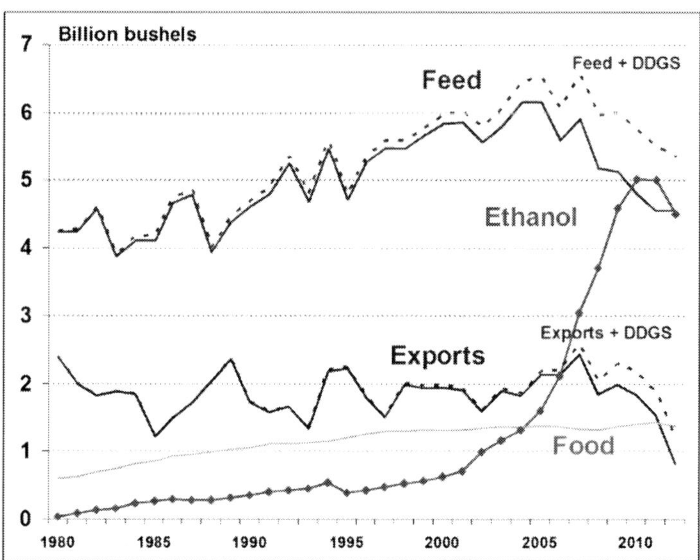

Source: USDA, Production, Supply, and Distribution (PSD) database, March 8, 2013.

Notes: Feed includes a residual category to balance USDA supply and demand estimates. The corn-to-ethanol production process generates a co-product, DDGS, which is a protein-rich animal feed. Both "Feed" and "Export" categories have been adjusted to include DDGS, as shown by the dotted lines.

Figure 4. Annual U.S. Corn Use by Major Activity, 1980 to 2012.

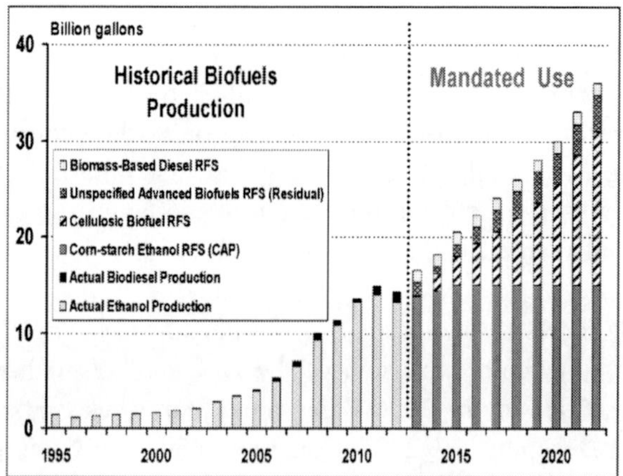

Sources: Actual ethanol production data for 1995-2012 is from Renewable Fuels Association; data for RFS2 mandates is from EISA (P.L. 110-140). Data includes proposed revision to RFS2 cellulosic mandate for 2013.

Figure 5. Renewable Fuels Standard (RFS2) vs. U.S. Ethanol Production Since 1995.

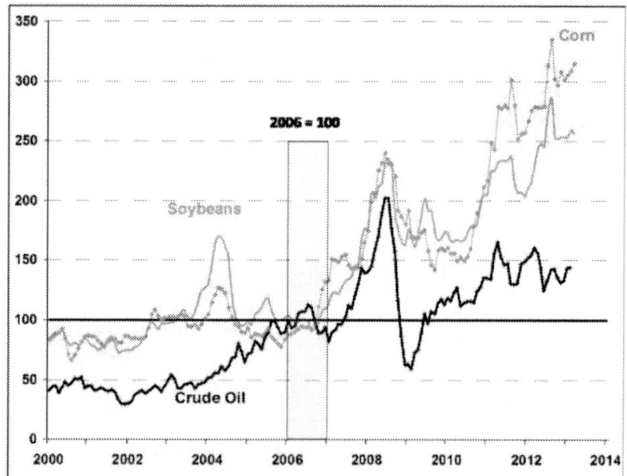

Sources: Corn and soybean prices are monthly average farm prices (MAFPs), National Agricultural Statistics Service (NASS), USDA; crude oil is the spot price, f.o.b., for West Texas Intermediate, Cushing, OK, EIA, DOE.

Notes: To facilitate comparison of relative price movements, the monthly prices have been converted by CRS to an index where the 12-month average for calendar 2006 has been set to 100.

Figure 6. Monthly Price Indexes for Corn, Soybeans, and Crude Oil, 2000 to 2013 (nominal monthly prices are indexed such that 2006 = 100).

The expanded RFS (referred to as RFS2) required the use of 9 bgals of biofuels in 2008 and expanded the mandate to 36 bgals annually in 2022. The new mandate had some provisos, foremost of which was that only 15 bgals of annual RFS-qualifying biofuels could be ethanol from corn starch. As a result, all increases in the RFS mandate from 2016 onward must be met by advanced biofuels (i.e., non-corn-starch biofuels) and no less than 16 bgals must be derived from cellulosic feedstock in 2022. In addition, the new mandate established by EISA carved out specific volume requirements for biomass-based diesel fuels.

Meanwhile, prices for many agricultural commodities—including nearly all major U.S. program crops—started a steady upward trend in late 2006. Then, in early 2007, the upward trend for commodity prices turned into a steep rise. By mid-2008 market prices for several agricultural commodities had reached record or near-record levels (Figure 6).[32] In particular, both corn and crude oil hit record high prices in both spot and futures markets, thus symbolizing the growing linkage between U.S. field crops and energy markets.[33]

The upward rise in the price of corn in 2007 and early 2008 sucked the profits out of the U.S. biofuels sector and put the brakes on new investment (Figure 3). It also fueled a "food-versus- fuel" debate about the potential for continued expansion in corn use for ethanol to have unintended consequences in other agricultural and environmental markets. While most economists and market analysts agreed that the dramatic price rise of 2008 was due to factors other than biofuels policy, they also are nearly universally agreed that the strong, steady growth in ethanol demand for corn has had an important and sustained upward price effect, not just on the price of corn, but in other agricultural markets including food, feed, fuel, and land.

By mid-2008, the commodity price rise had completely reversed itself and turned into a near free- fall, coinciding with the global financial crisis that broke in late 2008.[34] The extreme price volatility created many difficulties throughout the marketing chain for agricultural buyers and sellers. The experience of $7.00-per-bushel corn, albeit temporary, shattered the idea that biofuels were a panacea for solving the nation's energy security problems and left concerns about the potential for unintended consequences from future biofuels expansion.

2008 Farm Bill Reinforces Focus on Cellulosic Biofuels

The 2008 farm bill (Food, Conservation, and Energy Act of 2008; P.L. 110-246) extended and expanded many existing biofuels programs.[35] In particular, Title XV ("Trade and Tax Provisions") extended the biofuels tax incentives and the tariff on ethanol imports, although the tax credit for corn-starch ethanol was reduced to $0.45 per gallon. But in the wake of the commodity market price run-up of early 2008, the new farm bill also re-emphasized EISA's policy shift towards research and development of advanced and cellulosic bioenergy in an effort to avoid many of the unintended consequences of relying too heavily on major field crops as the principal biomass feedstock. In addition, it established a new tax credit of $1.01 per gallon for cellulosic biofuel.

Like the 2002 farm bill, it contained a distinct energy title (Title IX) that covers a wide range of energy and agricultural topics with extensive attention to biofuels, including corn starch-based ethanol, cellulosic ethanol, and biodiesel. Energy grants and loans are provided through initiatives such as the Bioenergy Program for Advanced Biofuels to promote the development of cellulosic biorefinery capacity. The Repowering Assistance Program supports increasing efficiencies in existing refineries. Programs such as the Rural Energy for America Program (REAP) assist rural communities and businesses in becoming more energy-efficient and self-sufficient, with an emphasis on small operations. Cellulosic feedstocks—for example, switchgrass and woody biomass—are given high priority both in research and funding. The Biomass Crop Assistance Program (BCAP), the Biorefinery Assistance Program, and the Forest Biomass for Energy Program provide support to develop alternative feedstock resources and the infrastructure to support the production, harvest, storage, and processing of cellulosic biomass feedstocks.

Title VII, the research title of the 2008 farm bill, contains numerous renewable-energy-related provisions that promote research, development, and demonstration of biomass-based renewable energy and biofuels. One of the major policy issues debated prior to the passage of the 2008 farm bill was the impact of the rapid, ethanol-driven expansion of U.S. corn production. This issue was made salient by the dramatic surge in commodity prices experienced in 2007 and early 2008. In partial consideration, the enacted bill requires reports on the economic impacts of ethanol production, reflecting concerns that the increasing share of corn production being used for ethanol contributed to high commodity prices and food price inflation.

However, funding authority for Title IX bioenergy programs was fairly limited—about $1 billion in mandatory funding and only slightly more than $100 million in discretionary funding was actually available during the life of the 2008 farm bill (FY2008-FY2012). In addition, all of the major Title IX bioenergy programs expired at the end of FY2012 and lacked baseline funding going forward. The 2008 farm bill (including Title IX) was extended through FY2013 by the American Taxpayer Relief Act (ATRA; P.L. 112-240).[36] However, all major bioenergy provisions of Title IX—with the exception of the Feedstock Flexibility Program for Bioenergy Producers— have no new mandatory funding in FY2013 under the ATRA farm bill extension.

Questions Emerge Concerning Rapid Biofuels Expansion

By 2009, more than half of all U.S. gasoline contained some ethanol (mostly blended at the 10% level or lower). However, national gasoline transportation fuel consumption peaked in 2007 at about 142.5 bgals and has been steadily declining—driven by a weak economy and improving passenger vehicle fuel economy. In 2010 U.S. ethanol consumption reached an estimated 12.9 billion gallons (bgals), which was blended into roughly 138 bgals of gasoline—this represents about 9.3 % of annual gasoline transportation demand on a volume basis.[37]

Meanwhile, robust economic growth in major global markets in 2010 and early 2011 (including China, India, Brazil, and other parts of Asia and the Middle East) reinvigorated international consumer demand and, when coupled with a weak U.S. dollar and events that occurred in international feed grain markets—drought in Russia, Kazakhstan, and the Ukraine in 2010, plus strong Chinese demand for corn and feedstuffs—contributed to record U.S. agricultural export values in 2010 and 2011 and helped to push commodity prices, especially corn, upward again.[38]

By 2010, U.S. ethanol production consumed 40% of the U.S. corn crop and surpassed corn-for- feed use for the first time in history (Figure 4). Combined strong demand from export markets and ethanol contributed to near historic low ending stock projections (relative to expected demand) for U.S. corn and soybean for 2010 and 2011.[39] These market conditions helped to spur another surge in agricultural commodity prices starting in mid-2010 (Figure 6), thus spreading the effects of rapidly expanding ethanol production and corn demand across several other sectors of the U.S. economy as well.

In addition to expanding domestic production of biofuels, there has been some interest in expanding imports of sugar-based ethanol—usually produced from sugar cane in Brazil—to help satisfy the RFS for advanced biofuels.[40] U.S. sugar-ethanol imports peaked at 660 million gallons in 2006 (including 434 million from Brazil). Market factors in 2010-2012—U.S. ethanol production approaching the "blend wall", high international sugar prices, lower-than-expected sugarcane output in Brazil, and a weak U.S. dollar—

resulted in the United States becoming a net exporter of ethanol during those years (Figure 7).[41]

Severe Drought Across Much of Corn Belt Slows Ethanol Industry

In early 2012, high market prices and nearly ideal springtime planting conditions across much of the United States led to substantial and extensive early corn planting. On June 12, 2012, USDA projected U.S. corn plantings of 95.9 million acres—the most since 1937. Normal weather patterns were expected to produce a record 2012 corn harvest of 14.8 billion bushels, which in turn would lead to a build-up in U.S. corn ending stocks in 2013 of nearly 2 billion bushels (up 111% year-to-year), and a 2012/2013 season-average corn price of $4.60/bushel (down 25%).[42] A record harvest and return to low corn prices were eagerly anticipated by both the ethanol and livestock industries.

However, in mid-June, an extensive swath of the Central and Southern Plains and much of the Corn Belt were hit by a combination of extreme heat and dryness that produced what was referred to as a "flash drought." By August 2012—just two months after its optimistic forecast of May— USDA had completely reversed its outlook from one of abundance to one of shortage. USDA lowered its forecast for U.S. corn production to 10.8 billion bushels (a 27% drop of 4 billion bushels from its May forecast), corn price projections were raised sharply to $8.20 per bushel (up 78%), and stocks of feed grains and soybeans were forecast to approach historic low levels relative to demand by the end of 2012/2013 crop year (i.e., at the end of summer 2013).[43]

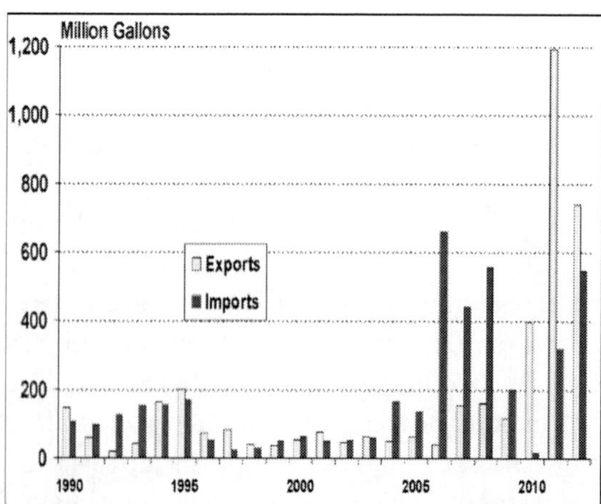

Source: U.S. Department of Commerce, U.S. Census Bureau, Foreign Trade Statistics.

Figure 7. Annual U.S. Ethanol Exports and Imports Since 1990.

Market prices for ethanol were not able to keep up with escalating production costs (primarily for corn) and negative production margins resulted in the idling of several ethanol facilities (Figure 3). As a result, U.S. ethanol production in 2012 declined to 13.3 billion gallons—the first decline in production since 1996, when then-record corn prices temporarily set back ethanol production. The outlook for low corn supplies until the 2013 corn harvest in the September-November period is expected to dampen ethanol production in 2013 as well, possibly reducing it below the 2012 level.[44]

RIN Prices Rise Sharply in Early 2013

Despite waning ethanol production, RFS mandates for biofuel use continued to grow in 2013 to 16.55 bgals of total biofuels, including 2.75 bgals advanced biofuels and a residual 13.8 bgals for corn ethanol. In contrast, national transportation consumption of gasoline-type fuels, which had hit its peak in 2007 at about 142.5 bgals, was projected at slightly under 131 bgals in 2013, with an implied ethanol blend wall of about 13 bgals.[45]

The price for renewable identification numbers (RINs)[46] for basic renewable ethanol (D6)—as reported from thinly traded markets[47]—soared from under $0.05 per gallon during most of 2012 to over $1.00 per gallon in early March 2013.[48] As a result, the RIN values for a fuel blender blending 1 million gallons of E10 (using 100,000 gallons of ethanol) in 2012 might have been $5,000 based on an average ethanol RIN price of about $0.05. The hypothetical value implied for that same volume at $1 per RIN would be $100,000.

The rapid RIN price increase is linked to the impending collision of the RFS mandates and the ethanol blend wall, which, without rapid expansion of the E15 or E85 markets, will likely require the use of accumulated RIN stocks for mandate compliance in 2013 and 2014.

Uncertainties Cloud Biofuels Future

In addition to the ethanol blend wall, the expanded RFS2 is likely to play a dominant role in the development of the U.S. biofuels sector, but with considerable uncertainty regarding spillover effects in other markets and on other important policy goals.[49] The rapid expansion of U.S. corn ethanol production and the concomitant dramatic rise in corn use for ethanol—USDA estimates that over 40% of both the 2011 and 2012 U.S. corn crops was used for ethanol production—has provoked questions about its long-run sustainability and the possibility of unintended consequences in other markets as well as for the environment.[50] Policymakers and the U.S. biofuels industry also are confronted by questions regarding the ability to meet the expanding RFS mandate for biofuels from non-corn sources such as cellulosic biomass materials, whose production capacity has been slow to develop,[51] or biomass-based diesel, which remains expensive to produce owing to the relatively high prices of its feedstocks.

It is widely believed that the ultimate success of the U.S. biofuels sector will depend on its ability to shift away from traditional row crops such as corn or soybeans for processing feedstock, and toward other, cheaper forms of biomass—such as prairie grass or algae—that do not compete with traditional food crops for land and other resources. Recent federal biofuels policies have attempted to assist this shift by focusing on the development of a cellulosic biofuels industry.[52] However, the speed of cellulosic biofuels development remains a major uncertainty, since new technologies must first emerge and be implemented on a commercial scale. The uncertainty surrounding the development of such new technologies and their commercial adaptation has been a major impediment to the flow of much needed private-sector investment funds into the cellulosic biofuels sector.

Ethanol Production Capacity Centered in Corn Belt

As of April 8, 2013, U.S. ethanol production was underway or planned in 210 plants located in 28 states based primarily around the central and western Corn Belt, where corn supplies are most plentiful (Table 2 and Figure 8). Existing U.S. ethanol plant capacity was estimated at 14.763 billion gallons per year (BGPY), with another 0.158 BGPY of capacity under construction (either as new plants or expansion of existing plants). Thus, total annual U.S. ethanol production capacity in existence or under construction was about 14.9 BGPY, well in excess of the 13.8 bgals RFS2 corn-starch ethanol residual quota for 2013 (Figure 2).

Iowa is by far the leading ethanol-producing state, with a 30% share of total U.S. output. The top six Corn Belt states of Iowa, Nebraska, Illinois, Minnesota, South Dakota, and Indiana account for nearly 75% of national production (Table 2). On a national level, actual operating capacity of 13.2 BGPY represents about 89% of nameplate capacity. This is because several states, including Nebraska, Minnesota, Indiana, Kansas, Ohio, and the "other" category of states, are operating substantially below their nameplate capacity, suggesting that poor industry profitability has been widespread across the country, primarily due to high feedstock cost and limited availability.

EVOLUTION OF THE U.S. BIODIESEL SECTOR

Biodiesel can be produced from any animal fat or vegetable oil (such as soybean oil or recycled cooking oil). Historically, most U.S. biodiesel was made from soybean oil. As a result, U.S. soybean producers and the American Soybean Association (ASA) are strong advocates for greater government support for biodiesel production. However, with the rise in soybean prices since 2007 (Figure 6), biodiesel producers have aggressively shifted to cheaper vegetable oils and animal fats (especially poultry fat), such that by 2011 nearly 44% of U.S. biodiesel production was estimated to be based on sources other than soybean oil.[53] In recent years, many ethanol production facilities have added technology to remove corn oil from distillers grains and solubles, thus generating an additional income stream to help offset depressed profit margins.[54] The corn oil produced by this "end-stream" technology is typically not suitable for the food industry. Instead, the main uses of this added corn oil has been as an energy supplement in livestock and poultry rations, and for biodiesel production.

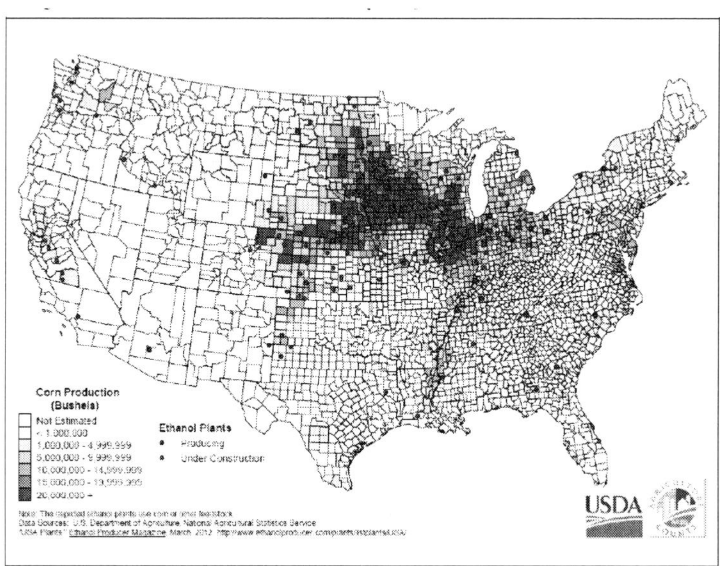

Source: USDA; U.S. corn production for 2011 compared with ethanol plant locations as of March 8; 2012; available at http://www.nass.usda.gov/ Charts_and_Maps/ Ethanol_Plants/U._S._Ethanol_Plants/index.asp.

Figure 8. U.S. Ethanol Production Capacity Is Centered on the Corn Belt.

Table 2. U.S. Ethanol Output and Production Capacity by State

Rank	State	# of Plants	Operating Production			Current Nameplate Capacity (MGPY)	Under Contr. or Expansion (MGPY)
			MGPY	% of output	Cumulative% output		
1	Iowa	41	3,903	30%	30%	3,908	—
2	Nebraska	27	1,509	11%	41%	1,822	—
3	Illinois	14	1,413	11%	52%	1,454	—
4	Minnesota	22	1,110	8%	60%	1,225	—
5	S. Dakota	15	1,016	8%	68%	1,016	—
6	Indiana	14	947	7%	75%	1,136	—
7	Wisconsin	9	504	4%	79%	504	5
8	Ohio	7	478	4%	82%	538	—
9	Kansas	13	386	3%	85%	507	45
10	N. Dakota	4	360	3%	88%	360	—
11	Michigan	5	268	2%	90%	268	—
12	Tennessee	2	225	2%	92%	225	—
13	Missouri	6	210	2%	93%	271	—
14	Texas	4	205	2%	95%	355	—
15	New York	2	164	1%	96%	164	—
	Others (13)	25	506	4%	100%	1,010	108
U.S. Total		**210**	**13,203**	**100%**		**14,763**	**158**

Source: Renewable Fuels Association as of April 8, 2013; state-level aggregations are by CRS and include several approximations of current plant operating levels.
Note: Output and production capacity data are in million gallons per year (MPGY).

According to the National Biodiesel Board (NBB), biodiesel is nontoxic, biodegradable, and essentially free of sulfur and aromatics. In addition, it works in any diesel engine with few or no modifications and offers similar fuel economy, horsepower, and torque, but with superior lubricity and important emission improvements over petroleum diesel.[55]

To date, biodiesel is used almost uniquely as a substitute for petroleum diesel transport fuel. Biodiesel delivers slightly less energy than petroleum diesel (about 92%); however, U.S. biodiesel consumption remains small relative to national diesel consumption levels. In 2012 (Figure 1), U.S. biodiesel consumption represented about 1.5% (in diesel-equivalent units) of national diesel transportation fuel use of about 46.8 billion gallons.[56]

Biodiesel is compatible with existing petroleum-based diesel vehicles and infrastructure (fuel tanks, retail pumps, delivery infrastructure etc.) such that biodiesel does not face a blend wall similar to ethanol. As a result, the potential blending pool for biodiesel is significantly larger than just the transportation diesel fuel market. Because biodiesel and diesel fuel are so similar, biodiesel can also be used for the same non-transportation activities—the two largest of which are home heating and power generation. In 2012, 53.2 billion gallons of diesel fuel were used for heating and power generation by residential, commercial, and industry, and by railroad and vessel traffic, bringing total U.S. diesel fuel use to nearly 106.7 billion gallons (including 46.8 billion gallons of transportation fuel use and 6.8 billion gallons of residual fuel oil).

Fuel blenders and consumers are very sensitive to price differences between biodiesel and petroleum-based diesel. The price relationship between vegetable oils and petroleum diesel is the key determinant of profitability in the biodiesel industry—about 7.5 pounds of vegetable oil are used in each gallon of biodiesel. Since late 2010, soybean oil prices have averaged over $0.50/lb. such that the vegetable oil feedstock component of biodiesel has cost over $3.75/gal. Additional processing and marketing costs likely push wholesale biodiesel prices into the $4.50/gal. to $5.00/gal. range compared with petroleum diesel wholesale prices of $3.05/gallon during that period. As a result, the biodiesel industry has depended on federal support—especially the production tax credit and the RFS for biomass-based diesel—for its economic survival.

Federal Programs Help Kick-Start U.S. Biodiesel Production

The U.S. biodiesel industry did not emerge until the late 1990s. In 1999, U.S. biodiesel production was still less than 1 million gallons. Bioenergy Program payments provided an initial impetus for biodiesel plant investments from 2001 through 2006. The American Jobs Creation Act of 2004 (P.L. 108-357) created the first ever federal biodiesel tax incentive—a federal excise tax and income tax credit of $1.00 for every gallon of agri-biodiesel (i.e., virgin vegetable oil and animal fat) that was used in blending with petroleum diesel; and a $0.50 credit for every gallon of non-agri-biodiesel (i.e., recycled oils such as yellow grease). The distinction between biodiesel from virgin and recycled oils was eventually removed (P.L. 110-343; October 3, 2008), and all biodiesel qualified for the credit of $1.00 per gal.

Starting in late 2005 through 2006, the U.S. biodiesel industry received a major economic boost from the same series of market and policy developments described for ethanol—i.e., high petroleum prices and low agricultural commodity prices.[57] Soybean oil prices were still relatively low priced during the 2000 through 2006 period, when they averaged $0.21/lb. (this compares with an average of nearly $0.44/lb. since 2007). The Energy Policy Act of 2005 extended the biodiesel tax credit and established a Small Agri-Biodiesel Producer Credit of $0.10 per gallon on the first 15 million gallons of biodiesel produced from plants with production capacity below 60 million gallons per year.

Biomass-based diesel (BBD) was not part of the initial biofuels RFS1 mandate under the Energy Policy Act of 2005, but was included as a distinct category in the RFS2 created under EISA of 2007. While most of this mandate is expected to be met using biodiesel, other fuels, including renewable diesel,[58] algae-based diesel, or cellulosic diesel, would also qualify.

Starting in mid-2007, the U.S. biodiesel industry suffered from unfavorable market conditions as prices for vegetable oil rose relative to diesel fuel (the monthly average wholesale price for soybean oil in Decatur, Illinois, hit $0.62/lb. in June 2008, implying a per-gallon cost of $4.65 for biodiesel). Most biodiesel plants continued to operate into 2008 in hopes of either higher diesel prices or lower vegetable oil prices, and the industry produced then-record output of an estimated 678 million gallons (Figure 9).[59] However, the financial crisis of late 2008 and the ensuing economic recession weakened demand for transportation fuel, and petroleum prices (including diesel fuel) fell sharply in the second half of 2008.

Starting in 2007 and 2008, U.S. biodiesel producers (relying heavily on the $1/gallon production tax credit) were able to take advantage of a favorable price relationship vis-à-vis the European Union (EU)—which also had domestic policies that encouraged biodiesel consumption—and profitably exported substantial volumes of U.S.-produced biodiesel to the EU. As a result, U.S. biodiesel exports soared to a record 677 million gallons in 2008.

However, in March 2009, the EU imposed anti-dumping and countervailing duty tariffs on imports of U.S. biodiesel that effectively shut down U.S. biodiesel exports to the EU and cut in half a major supply outlet for U.S. biodiesel producers (Figure 10).[61]

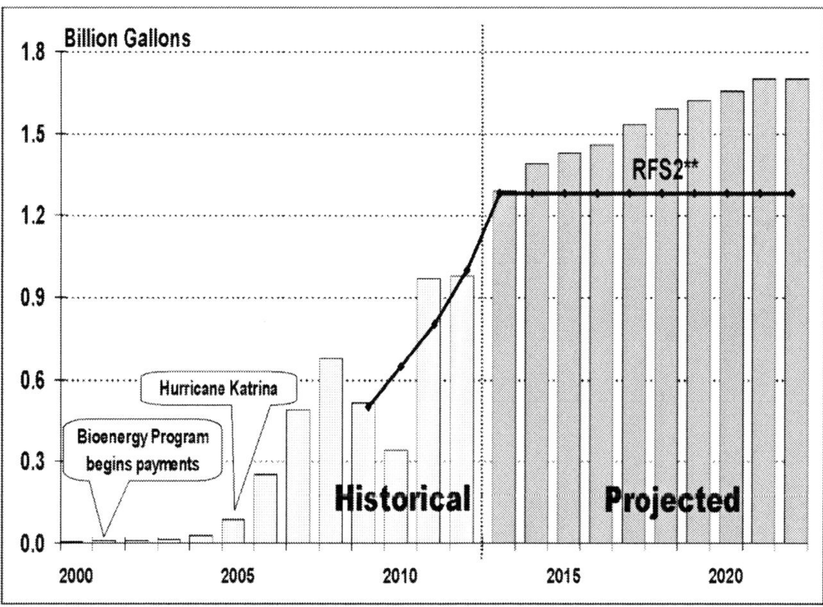

Source: Data for 1999-2012, Energy Information Agency, DOE. Projections for 2013-2022 are from FAPRI, *FAPRI-MU Biofuel Baseline*, FAPRI-MU Report #02-13, March 2013. FAPRI projections assume that market conditions, driven in part by the RFS for advanced biofuel, result in BBD consumption above the RFS for BBD.

Notes: RFS2** shown in the chart represents the RFS for BBD. Although the RFS2 mandate for biodiesel was to begin in 2009, implementation rules were not available until February 2010. As a result, the RFS2 mandate for 2009 of 500 million gallons was combined with the 2010 mandate of 650 million gallons for a one-time mandate of 1.15 billion gallons in 2010. In 2011, the mandate returned to its original trajectory of 800 million gallons, rising to 1 billion gallons in 2012. Starting in 2013, EPA is directed to establish the BBD RFS at no less than 1 billion gallons through a future rulemaking. In its 2013 RFS proposal, EPA proposed a BBD RFS of 1.28 billion gallons.[60] FAPRI assumes that it remains at that level through FY2022.

Figure 9. Annual U.S. Bio-Based Diesel (BBD) Production 1999 to 2022.

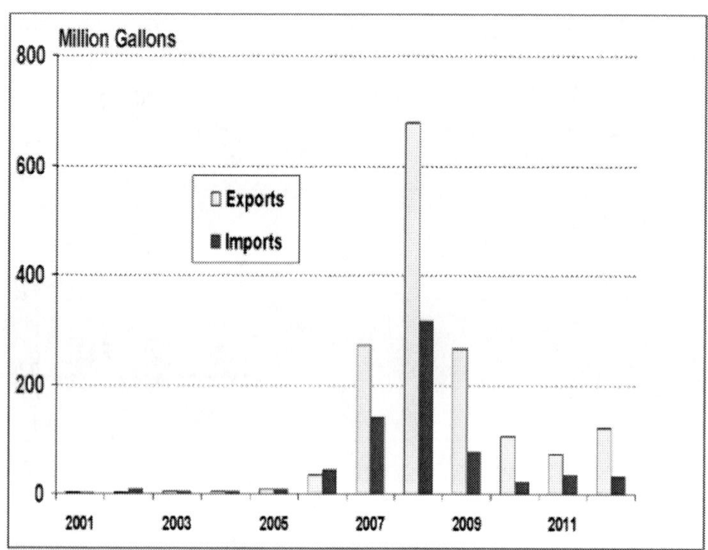

Source: EIA, *Monthly Energy Review*, March 2013, Table 10.4 "Biodiesel Overview."

Figure 10. Annual U.S. Biodiesel Exports and Imports Since 2001.

As a result, the U.S. biodiesel industry experienced several bankruptcies and some loss of capacity during 2009. U.S. biodiesel production in 2009 fell to 516 million gallons, down 24% from 2008.[62] The unfavorable economic conditions for biodiesel production extended into 2010 and were made worse by the expiration of the biodiesel tax credit at the end of 2009. The tax credit was eventually renewed on December 17, 2010 (P.L. 111-312), and made available retroactively to all 2010 biodiesel production; however, the extended delay and poor market conditions contributed to substantially reduced U.S. biodiesel production of 343 million gallons in 2010. During 2010, the U.S. biodiesel industry saw 52 out of 170 operating plants stop operations while many others scaled back on production.[63] The renewal of the tax credit and the expanded RFS2 biodiesel usage mandate of 800 million gallons in 2011 revived the industry and spurred record production of 967 million gallons in 2011 (Figure 9).

Once again both the biodiesel tax credit ($1.00/gallon) and the small agri-biodiesel producer credit ($0.10/gallon on the first 15 million gallons) expired at the end of 2011, but were extended through 2013 by P.L. 112-240, which retroactively applied the extension to fuel produced in 2012. In addition to the retroactive tax credit, biodiesel production in 2012 was supported by the RFS2 biodiesel mandate, which grew to 1 billion gallons in 2012. U.S. biodiesel production eclipsed the previous year's record with an output of 969 million gallons in 2012.

Two factors are expected to support biodiesel production at or above 1.28 billion gallons starting in 2013 and going forward: first, the RFS2 biodiesel mandate for 2013 has been proposed at 1.28 billion gallons by EPA; second, the RFS2 for advanced biofuels (for which biodiesel is a qualifying fuel) grows even faster, with 2.75 billion gallons in 2013 rising to 21 billion gallons by 2022. Although cellulosic biofuel was originally envisioned to fill most of the advanced biofuel mandate, slow progress in commercial production to date suggests that biodiesel may be used to meet at least a portion of the advanced biofuel mandate in the future. If this projected outcome were to be realized, it would likely have a profound impact on vegetable oil markets, as biodiesel production would be expected to consume an increasingly larger share of available supplies.[64]

Biodiesel Production Capacity Spreads Nationwide

As mentioned earlier, the primary feedstock for biodiesel includes both vegetable oils and animal fats, both of which are produced over a greater geographic area than corn. As a result, biodiesel plants are more widely dispersed across the United States than are ethanol plants (Table 3). As of January 2013, there were 110 companies in the United States with the potential to produce biodiesel commercially that were either in operation or idled, with total annual production capacity (within the oleo-chemical industry) of 2.1 billion gallons per year. Because many of these plants also can produce other products such as cosmetics, estimated total capacity (and capacity for expansion) is far greater than actual biodiesel production.

The unfavorable economic conditions of 2009 and 2010, coupled with the delays in extending the biodiesel tax credit first in 2010 and then again in 2012, and finally the run-up in soybean and product prices in 2011 and 2012, all contributed to a substantial shake-up in the biodiesel industry. Many plants situated in the heart of corn and soybean country dropped out of business, while new plants sprang up in locations near alternate vegetable or animal oil sources. As a result, the U.S. biodiesel industry is more diversified and less centralized than the ethanol industry. Unlike ethanol, where the top six producing states account for 75% of national capacity, the top six biodiesel-producing states achieve only a 58% share, thus demonstrating the more widespread nature of U.S. biodiesel production capacity.

U.S. TRANSPORTATION FUEL INFRASTRUCTURE

A key determinant of the demand for biofuels as a transportation fuel is the size and fuel economy of the U.S. vehicle fleet, and the adequacy of the infrastructure (e.g., pipelines, storage tanks, service pumps) that delivers transportation fuel to consumers at the retail level. According to the Department of Energy (DOE), 73% of U.S. transportation fuel is consumed as gasoline or gasoline blends (Figure 1), with the remainder consumed as diesel fuel. Gasoline blends and diesel fuel, for the most part, require different infrastructure for delivery to the retail market. In addition, vehicle motors are designed to operate with either gasoline or diesel, but not both.

U.S. Vehicle Fleet

The U.S. Department of Transportation (DOT) estimated that there were 250.2 million registered passenger vehicles (including trucks, buses, and motorcycles) in the United States in 2011, down slightly from 254.2 million in 2009.[65] Included in the fleet of passenger vehicles are more than 14 million flex-fuel vehicles (FFVs), which are capable of operating on the standard 10% ethanol and 90% gasoline (E10) blends as well as higher ethanol blends up to 85% ethanol and 15% gasoline (E85).[66]

Gasoline-Blend Infrastructure Issues

Because of its physical properties, pure ethanol cannot be used in the same infrastructure used to deliver retail gasoline. Nor can ethanol be used in standard automobile engines at high blend ratios, because ethanol tends to make the engine run at a higher temperature than standard reformulated gasoline. In addition, the presence of ethanol can be corrosive on rubber and plastic parts in the car engine. In contrast, biodiesel is very similar in nature to petroleum diesel and does not have the same infrastructure limitations.

The Blend Wall and Higher-Level Ethanol Blends

Prior to October 2010, the amount of ethanol that could be blended in gasoline for use in standard vehicle motors without modification was limited to 10% by volume (E10), by guidance developed by the EPA under the Clean Air Act, and certification procedures for fuel-dispensing equipment. In addition, most vehicle warranties did not cover any motor damage resulting from use of ethanol blends above 10%. In the past, only flex-fuel vehicles (FFVs) have been capable of using higher ethanol blends.

As a result, this 10% blend has represented an upper bound (sometimes referred to as the "blend wall") to the amount of ethanol that can be introduced into the gasoline pool.[67] If most or all gasoline in the country contained 10% ethanol, this would allow only for roughly 13 billion gallons, far less than the RFS mandates for 2013 onward.

For ethanol consumption to exceed the so-called blend wall and meet the RFS mandates, increased consumption at higher blending ratios is needed. For example, raising the blending limit from 10% to a higher ratio such as 15% or 20% would immediately expand the "blend wall" to somewhere in the range of 20 billion to 27 billion gallons. The U.S. ethanol industry is a strong proponent of raising the blending ratio.

Table 3. U.S. Biodiesel Production Capacity Partial Estimate as of January 2013

Rank	State	# of Plants	Production Capacity (MGY)	% of Output	Cumulative % output
1	Texas	11	408	20%	20%
2	Iowa	8	250	12%	32%
3	Missouri	8	170	8%	40%
4	Illinois	5	166	8%	48%
5	Washington	4	109	5%	53%
6	Minnesota	4	107	5%	58%
7	Mississippi	3	105	5%	63%
8	Indiana	2	104	5%	68%
9	Pennsylvania	6	90	4%	72%
10	Arkansas	3	85	4%	76%
11	N. Dakota	1	85	4%	80%
12	Kentucky	5	68	3%	84%
13	Ohio	3	67	3%	87%
14	California	9	57	3%	90%
15	Alabama	2	49	2%	92%
	Others (22)	36	168	8%	100%
U.S. Total		**110**	**2,086**	**100%**	

Source: U.S. EIA, "Table 4. Biodiesel Producers and Production Capacity by State, January 2013," *Monthly Biodiesel Production Report*, March 28, 2013.

The blend wall problem is made more acute by substantial revisions in EIA's projections of U.S. transportation fuel consumption rates since the RFS was first passed into law in 2007 (Figure 11). At that time, EIA estimated that U.S. transportation consumers were using about 145 billion gallons of gasoline (including ethanol) per year, but that consumption would grow strongly to 176 billion gallons of gasoline by 2022—as a result, RFS mandated biofuels would represent about 19% of annual gasoline consumption. By 2013, EIA had substantially lowered its fuel consumption outlook—partly due to sustained high petroleum prices, the prolonged effects of the 2008 financial crisis on consumer incomes, and significantly higher fuel economy standards on new vehicles. Instead of growth, EIA projects gasoline consumption to fall to about 120 billion gallons by 2022, thus causing the RFS mandate's share of the gasoline transportation fuel market to grow to nearly 20% of annual consumption (in gasoline-equivalent gallons).[68]

EPA Ruling on the Ethanol-to-Gasoline Blending Limit: 10% vs. 15%

On March 6, 2009, Growth Energy (on behalf of 52 U.S. ethanol producers) applied to the EPA for a waiver from the then-current Clean Air Act E10 limit and an increase in the maximum allowable concentration to 15% (E15). After substantial vehicle testing, the EPA issued, first a partial waiver (October 2010) for gasoline that contains up to a 15% ethanol blend (E15) for use in model year 2007 or newer passenger vehicles (including cars, SUVs, and light pickup trucks).[69] Then after further testing, on January 21, 2011, EPA expanded the eligible passenger vehicle pool to include model years 2001 through 2006.[70]

However, EPA also announced that no waiver would be granted for E15 use in model year 2000 and older light-duty motor vehicles, as well as in any motorcycles, heavy duty vehicles, or non- road engines. This later restriction opens up the possibility of "mis-fueling"—that is, using higher ethanol blends in vehicles not appropriate for the EPA 15% blend waiver.[71] According to the Renewable Fuel Association (RFA), the approval of E15 use in model year 2001 and newer passenger vehicles covered 62% of passenger vehicles on U.S. roads at the end of 2010.[72]

These EPA rulings would appear to have expanded the eligible vehicle pool for ethanol blends greater than 10%. However, two factors prevent a blend wall expansion to 15%. First, U.S. automakers have not yet extended vehicle warranties to cover any motor damage resulting from use of ethanol blends above 10%. Second, the fact that a portion of currently active passenger vehicles are not eligible for E15—i.e., model year 2000 or older—both limits ethanol retail delivery opportunities and raises the cost of delivery, thus inhibiting retailer adoption.

Alternate Options to the Blend Wall

Two additional options to resolving this bottleneck exist, but appear to be long-run alternatives. The first is to increase the use of ethanol in flex-fuel vehicles (FFVs) at ethanol-to-gasoline blend ratios as high as E85. However, increased E85 use would involve substantial infrastructure development, particularly in the number of designated storage tanks and E85 retail pumps, as well as a further expansion of the FFV fleet to absorb larger volumes of ethanol.

According to the Renewable Fuels Association (RFA), more than 14 million FFVs were on the roads in 2012, representing over 5% of U.S. passenger vehicles. However, not all FFV owners have access to (or choose to use) E85 retail pumps. As of early 2013, over 3,000 retail stations in the United States offered E85 (2% out of 142,000 stations).[73] Most E85 fueling stations are concentrated in the midwestern states near the current ethanol production heartland (Figure 12).

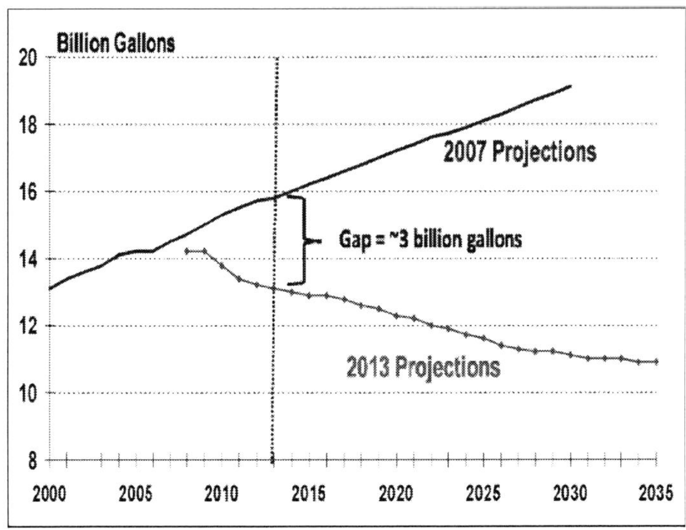

Source: Calculations are by CRS based on data from EIA, DOE, *Annual Energy Review* 2007 and *Annual Energy Review* 2013.

Notes: The blend wall is calculated as a simple 10% share of projections for U.S. gasoline consumption.

Figure 11. Ethanol Blend Wall Projections, 2007 vs. 2013.

In addition, at blend ratios above 10%, ethanol must compete directly with gasoline as a transportation fuel. For ethanol to operate primarily as a gasoline substitute, it must be priced competitively with gasoline on an energy-content or miles-per-gallon basis.

A second alternative is to expand use of processing technologies at the biofuel plant to produce biofuels in a “drop-in” form (e.g., butanol) that can be used by existing petroleum-based distribution and storage infrastructure and the current fleet of U.S. vehicles. However, more infrastructure-friendly biofuels generally require more processing than ethanol and are therefore more expensive to produce.

FEDERAL PROGRAMS THAT SUPPORT BIOFUELS

Federal Biofuels Policies Have Encouraged Rapid Growth ...

Federal biofuels programs have proven critical to the economic success of the U.S. biofuels industry, primarily ethanol and biodiesel, whose output has grown rapidly in recent years. Initially, federal biofuels policies were developed to help kick-start the biofuels industry during its early development, when neither production capacity nor a market for the finished product were widely available. Federal policy played a key role in underwriting the initial investments in biofuels production capacity as well as in helping to close the price gap between biofuels and cheaper petroleum fuels.

During the rapid growth period of 2006-2011, U.S. biofuels production and supporting federal budget outlays grew concomitantly. Federal support for biofuels production peaked in 2011, when an estimated $7.7 billion of direct support—including tax credit expenditures ($7.3 billion) and 2008 farm bill Title IX outlays (approximately $300 million)—was incurred.[74] Federal outlays in 2012 are estimated sharply lower, at about $1.3 billion, due to the expiration of several biofuels tax credits.

... And Conflicting Viewpoints

The trade-offs between benefits to farm and rural economies, as opposed to large federal budget costs and the potential for unintended consequences, have led to emergence of both proponents and critics of the government subsidies and mandates that underwrite biofuels production. Oversight and implementation of federal biofuels policies is spread across several government agencies, but the primary responsibility lies with EPA, USDA, and DOE. As the number, complexity, and budgetary implications of federal biofuels policies have grown, so too has the number of proponents and critics.

Proponents of government support for agriculture-based biofuels production have cited national energy security, reductions in greenhouse gas emissions, and raising domestic demand for U.S.- produced farm products as viable justifications. In many cases, biofuels are more environmentally friendly (in terms of emissions of toxins, volatile organic compounds, and greenhouse gases) than petroleum products. In addition, proponents argue that rural, agriculture-based energy production can enhance rural incomes and expand employment opportunities, while encouraging greater value-added for U.S. agricultural commodities.[75]

In contrast, critics argue that, in the absence of subsidies, current biofuels production strategies can only be economically competitive with existing fossil fuels at much higher petroleum prices, or if significant improvements in existing technologies are made or new technologies are developed.[76] Until such technological breakthroughs are achieved, critics contend that the subsidies distort energy market incentives and divert research funds from the development of other renewable energy sources, such as solar or geothermal, that offer potentially cleaner, more bountiful alternatives. Still others question the rationale behind policies that promote biofuels for energy security. These critics question whether the United States could ever produce sufficient feedstock of starches, sugars, or vegetable oils to permit biofuels production to meaningfully offset petroleum imports.[77] Critics from the petroleum industry argue against the economic costs associated with the imposition of biofuels blending requirements.[78] Finally, some (particularly environmental watchdog groups) argue that the focus on development of alternative energy sources undermines efforts for greater conservation to reduce energy waste.

Many biofuels-related policy debates occur along geographic lines. For example, Midwest corn- and ethanol-producing states are major proponents of federal policy support, whereas many residents of the East and West Coast urban states perceive expensive biofuel usage mandates as being forced upon them while their access to cheaper Brazilian sugar-cane ethanol was, for many years, limited by an import tariff. Another source of biofuels policy conflict has emerged between the major users of corn. Livestock producers have seen their feed costs escalate with the growth in biofuels corn demand and are highly critical of further federal biofuels support.

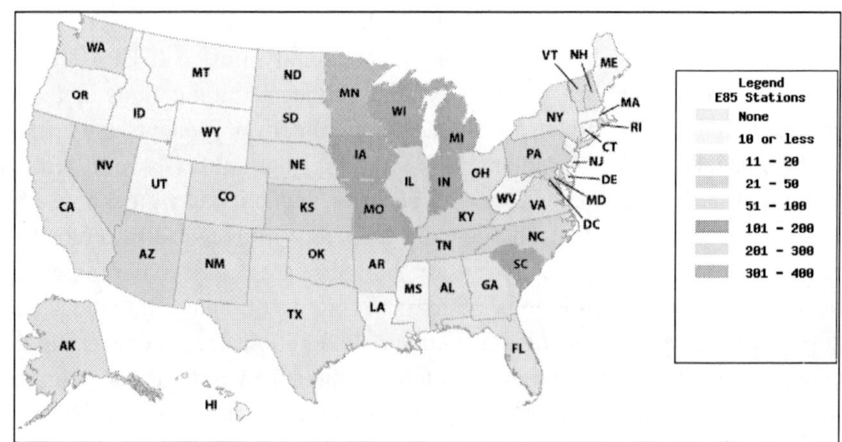

Source: U.S. DOE, Alternative Fuels and Advanced Vehicles Data Center, November 2010, available at http://www.afdc.energy.gov/afdc/ethanol/ethanol_locations.html.

Figure 12. E85 Refueling Locations by State.

Federal Biofuels Programs Described

Most of the biofuels policies developed and funded by Congress are subject to oversight and periodic reauthorization.[79] For most of the past three decades, three types of federal programs have provided the core support for the U.S. biofuels industry: blending and production tax credits to lower the cost of biofuels to end users, an import tariff to protect domestic ethanol from cheaper foreign-produced ethanol, and volume-specific usage mandates to guarantee a market for biofuels irrespective of their cost. In addition, the biofuels industry has been supported by several indirect policies in the form of research grants to stimulate the development of new technologies, and grants, loans, and loan guarantees to facilitate the development of biofuels feedstocks as well as market and distribution infrastructure.

Tax Credits

Various tax credits and other incentives have been available for the production, blending, and/or sale of biofuels and biofuel blends (Table 4). Tax credits vary by the type of fuel and the size of the producer. Because of their budgetary cost, the tax credits are rarely extended for more than a year or two at a time. As a result, they routinely require congressional action to be extended. On December 31, 2011, most biofuels blending and production tax credits expired, with the exception of the cellulosic biofuels production tax credit, which was set to expire at the end of 2012. The American Tax Payer Relief Act of 2012 (P.L. 112-240) extended both the producer and small producer tax credits for biodiesel, renewable diesel, and cellulosic biofuels through 2013 and retroactively for 2012.

Import Tariff on Foreign-Produced Ethanol

Prior to 2012, most imported ethanol was subject to a most-favored-nation duty set of $0.54 per gallon of ethanol (for fuel use) and a 2.5% ad valorem tariff. The stated goal of the import tariff was to offset the ethanol blending tax credit which was also available for foreign-produced ethanol. However, the fixed $0.54-per-gallon most favored-nation duty (identified by 9901.00.50 and 9901.00.52 of the Harmonized Tariff System (HTS)) expired on December 31, 2011. The 2.5% ad valorem tariff (2207.10.60 of the HTS) does not expire but is permanent until or unless the HTS code itself is changed. In most years the tariff was a significant barrier to direct imports of Brazilian sugarcane ethanol. However, some Brazilian ethanol could be brought into the United States duty-free if it was dehydrated (reprocessed) in Caribbean Basin Initiative (CBI) countries.[80] Up to 7% of the U.S. ethanol market could be supplied duty-free in this fashion; historically, however, ethanol dehydrated in CBI countries has only represented about 2% of the total U.S. market.

The Renewable Fuel Standard (RFS)[81]

As described earlier, the RFS requires the blending of renewable fuels (including ethanol and biodiesel) in U.S. transportation fuel.[82] The RFS is administered by EPA. Under the RFS, fuel blenders are required to blend an increasing amount of renewable fuel in the national transportation fuel supply. This requirement increases annually from 9 billion gallons (bgals) in 2008 to 36 bgals in 2022, of which only 15 bgals can be ethanol from corn starch. The remaining 21 bgals are to be so-called "advanced biofuels"—fuels produced from non-corn-starch feedstocks—of which 16 bgals are to be from cellulosic biofuels, 1 bgals from biomass-based diesel, and 4 bgals from other biofuels (most likely imported sugar-cane ethanol from Brazil). Qualifying biofuels must meet explicit criteria on lifecycle greenhouse gas (GHG) emissions[83] and feedstock production pathways (including restrictions on the land on which feedstocks are produced, feedstock production methods, and the biofuels plant processing technology).

Table 4. Federal Tax Credits Available for Qualifying Biofuels

Biofuel	Tax Credit: $/gallon	Details	Expiration Date
Volumetric Ethanol Excise Tax Credit (VEETC)	$0.45	Available in unlimited amount to all qualifying biofuels.	Expired Dec. 31, 2011
Small Ethanol Producer Credit	$0.10	Available on the first 15 million gallons (mgal) of any producer with production capacity below 60 mgal.	Expired Dec. 31, 2011
Biodiesel Tax Credit	$1.00	Available in unlimited amount to all qualifying biodiesel.	Dec. 31, 2013[a]
Small Agri-Biodiesel Producer Credit	$0.10	Available on the first 15 mgal of any producer with production capacity below 60 mgal.	Dec. 31, 2013[a]
Renewable Diesel Tax Credit	$1.00	Available in unlimited amount to all qualifying biodiesel.	Dec. 31, 2013[a]
Credit for Production of Cellulosic and Algae-Based Biofuel[b]	$1.01	Available in unlimited amount to all qualifying biofuels.	Dec. 31, 2013[a]

Source: CRS Report R42566, *Alternative Fuel and Advanced Vehicle Technology Incentives: A Summary of Federal Programs.*

[a] The tax credit originally expired at the end of 2009 and was not extended until the passage of P.L. 111-312, which retroactively applied the extension to fuel produced in 2010. The tax credit also expired at the end of 2011 and was extended through 2013 by P.L. 112-240, which retroactively applied the extension to fuel produced in 2012.

[b] P.L. 112-240, amended the credit to included non-cellulosic fuel produced from algae feedstocks.

Other Indirect Federal Policies

Several additional biofuels programs have been created to provide various grants, loans, and loan guarantees in support of research and development of related technology, as well as support for biofuels infrastructure development. Many of these programs reside in the energy title (Title IX) of the 2008 farm bill (P.L. 110-246).[84] Federal programs also require federal agencies to give preference to bio-based products in purchasing fuels and other supplies. Cellulosic plant investment is further facilitated by a special depreciation allowance created under the Tax Relief and Health Care Act of 2006 (P.L. 109-432).[85] Also, several states have their own incentives, regulations, and programs in support of renewable fuel research, production, and use that supplement or exceed federal incentives.[86]

In addition to direct and indirect biofuels policies, the U.S. biofuels industry benefits from U.S. farm programs in the form of price and income support programs (i.e., marketing loan benefits and the counter-cyclical payment program) and risk-reducing farm programs (e.g., Acreage Crop Revenue Election (ACRE), Supplemental Revenue Assistance Payments (SURE), federal crop insurance, and disaster assistance), which encourage greater production and lower prices than would occur in the absence of federal programs in a free-market equilibrium.[87] As a result, agricultural feedstocks are both lower-priced and more abundant than without federal farm programs. This helps lower production costs for the U.S. biofuels sector, and makes U.S. biofuels more competitive with foreign-produced biofuels.

CURRENT BIOFUELS POLICY ISSUES

Most of the federal biofuels tax credit provisions, as well as the import tariff on foreign-produced ethanol, have short legislative lives and require frequent extension. The primary energy-related issue for the next farm bill is the expiration of program authority at the end of FY2013 and the current lack of mandatory funding going forward for all major energy-related provisions of Title IX.[88] In addition, the appearance of substantial redundancy across renewable energy programs at USDA and DOE, the slow development of the U.S. cellulosic biofuels sector, and concerns about the emerging spillover effects of increasing corn use for ethanol production are issues that are likely to emerge during the next farm bill debate.

Pending Congressional Actions

2008 Farm Bill Expiration

Many provisions of the 2008 farm bill expired at the end of FY2012, but were extended through FY2013 by the American Taxpayer Relief Act (ATRA; P.L. 112-240).[89] Authority for Title IX biofuels policy provisions contained in the 2008 farm bill (P.L. 110-246) also were extended through FY2013, and are expected to be reviewed as part of the next farm bill debate.[90] However, all major bioenergy provisions of Title IX—with the exception of the Feedstock Flexibility Program for Bioenergy Producers—have no new mandatory funding in FY2013 under the ATRA farm bill extension. Although most of the bioenergy programs are reauthorized for FY2013, their mandatory funding expired at the end of FY2012. If policymakers want to continue these programs under either the 2008 farm bill extension or in the next farm bill, they will need to pay for the program with offsets.

The 2008 farm bill authorized $1.1 billion in mandatory funding for energy programs, including $320 million for the Biorefinery Assistance Program, $300 million for the Bioenergy Program for Advanced Biofuels, and $255 million for the Rural Energy for America Program (REAP). The Biomass Crop Assistance Program (BCAP) was authorized to receive such sums as necessary (i.e., funding is open-ended and depends on program participation), although Congress eventually put limits on mandatory funding of $552 million in FY2010, $112 million in FY2011, and $17 million in FY2012. None of the major farm-bill energy programs have baseline funding after FY2012. As a result, the federal budget rules require new revenues or offsetting cuts in order to extend them beyond FY2012.

Cellulosic Biofuels Tax Credit

While most ethanol tax credits and the import duty on foreign fuel ethanol expired on December 31, 2011, the cellulosic biofuel tax credit and the various biodiesel tax credits do not expire until December 31, 2013. Both the cellulosic biofuels and biodiesel industries can be expected to lobby actively for extension of their tax credits. However, a tight federal budget combined with lack of progress in developing commercial production of cellulosic biofuels are likely to work against an extension. At $1.00 per gallon, the biodiesel tax credit is projected to cost at least $1.28 billion in tax expenditures in 2012, whereas the cellulosic biofuels tax credit is projected to cost about $14 million.

Cellulosic Biofuels Feedstock Program: BCAP

Investors have been slow to invest in what so far is a commercially unproven technology—the conversion of cellulosic biomass to biofuels. Development of the cellulosic biofuels industry hinges on the effective use of new feedstocks. The Biomass Crop Assistance Program (BCAP) was created under the 2008 farm bill to facilitate the development of those new feedstocks and kick-start the cellulosic biofuels industry.[91] BCAP (via USDA's CCC) provides financial assistance in two forms: (1) to support the establishment and production of eligible crops for conversion to bioenergy in selected areas, and (2) to assist agricultural and forest land owners and operators with collection, harvest, storage, and transportation (CHST) of eligible material for use in a biomass conversion facility.

While BCAP is in the early stages of implementation, concerns regarding eligibility, funding, and sustainability continue to be discussed. These issues could shape future congressional action on the program in the context of budgetary measures and possible reauthorization in the next farm bill. In particular, BCAP does not include "baseline" budget spending beyond FY2012. Based on current budgetary requirements, the authorizing committees could potentially need to secure offset funding if BCAP were to be reauthorized in the next farm bill. This could prove difficult given tight budgetary constraints and the more recent and higher projections of the program's cost compared to its initial cost estimates.

Proposed Biofuels-Related Bills in the 113th Congress

The current federal biofuels programs continue to inspire strong sentiments from both advocates and detractors. Several Members of Congress have introduced bills that would either strengthen or reduce (and even eliminate) certain features of current programs.

Pending EPA Actions

As administrator of the RFS program, the EPA is responsible for identifying renewable fuel production pathways and pathway components that can be used in producing qualifying renewable fuel under the RFS program. The EPA is also responsible for announcing the RFS mandate levels for each year based on an evaluation and determination of the estimated production capacity (both domestic and international) of the various biofuels types. If it appears that the production capacity will be insufficient for a particular biofuel category—e.g., cellulosic biofuels—then EPA may announce a waiver of the original statutory RFS mandate for that category (and possibly other nested categories) to a reduced level. In addition, EPA may entertain RFS waiver petitions regarding potential economic hardship related to meeting a particular RFS mandate category.

Table 5. Selected Biofuels-Related Bills in the 113th Congress

Bill Number	Bill Name	Sponsor	Action
H.R. 550 S. 251	Phantom Fuel Reform Act of 2013	Rep. Gregg Harper Sen. Flake	To amend the RFS to require the cellulosic biofuel requirement to be based on actual production for the Jan.-Oct. period of the preceding year, pro-rated to an annual basis.
H.R. 596	Public Lands Renewable Energy Development Act of 2013	Rep. Paul Gosar	To promote the development of renewable energy on public lands.
H.R. 796	Amendment to the Clean Air Act	Rep. Sensenbrenner	To limit the cellulosic RFS mandate to be not more than 5% or 1 million gallons (whichever is greater) more than the total volume of cellulosic biofuel that was commercially available for the most recent calendar year.
H.R. 875	untitled	Rep. Sensenbrenner	To provide for a comprehensive assessment of the scientific and technical research on the implications of the use of mid-level ethanol blends (e.g., E15).
H.R. 979	Forest Products Fairness Act of 2013	Rep. Thompson	To modify the definition of the term `biobased product' to more broadly include forest products.
H.R. 1214	Domestic Fuels Protection Act of 2013	Rep. Shimkus	To provide liability protection for claims based on the design, manufacture, sale, offer for sale, introduction into commerce, or use of certain fuels and fuel additives (e.g., E15).

Bill Number	Bill Name	Sponsor	Action
H.R. 1273	Rural Energy Improvement Act	Rep. Welch	To reauthorize and improve the Rural Energy for America Program (REAP).
H.R. 1461	RFS Elimination Act	Rep. Goodlatte	To repeal the RFS program of the EPA.
H.R. 1462 S. 344	RFS Reform Act of 2013	Rep. Goodlatte Sen. Wicker	To prohibit the EPA from approving the introduction into commerce of gasoline that contains greater than 10%-volume ethanol
H.R. 1469	Leave Ethanol Volumes at Existing Levels (LEVEL) Act	Rep. Burgess	To limit expansion of RFS biofuel mandates, to prohibit authorization of ethanol blends greater than 10%.
H.R. 1482	RFS Amendments Act	Rep. Womak	To eliminate corn ethanol requirements under the RFS program
S. 289	Freedom Fuels Act of 2013	Sen. Baucus	To authorize long-term contracts for the procurement of certain liquid transportation fuels for the Dept. of Defense

Source: Legislative Information System of the U.S. Congress.

Notes: This is not meant to serve as a comprehensive list of all energy-related bills, but instead represents a selection of bills deemed (by CRS) most relevant to federal biofuels programs and policies.

Waiver of Mandated Use Requirements

The RFS mandates the use of over 16.55 bgals of biofuels in 2013. The mandate grows to 20.5 bgals of biofuels use by 2015. By 2022, 36 bgals of biofuels must be consumed under the RFS. Each year EPA must review the likelihood of outyear biofuel production meeting or failing to meet required RFS usage levels, and adjust the mandates accordingly. EPA's biofuels standards for each upcoming year are announced on a preliminary basis in the spring of the preceding year, when EPA issues a notice of proposed rulemaking, and on a final basis by November 30 of the preceding year, when EPA issues a final rule.[92]

The EPA has already waived the original RFS2 mandate for cellulosic biofuels for each of the first three years (2010, 2011, and 2012) and has proposed waiving it for a fourth year (2013). The likelihood of future EPA waivers could deter capital investments in the sector and make future waivers become a self-fulfilling prophecy. The likelihood of meeting RFS mandates for traditional biofuels hinges both on the "blend wall" and on the slow emergence of a national infrastructure needed to facilitate the distribution and use of the growing mandated biofuel volumes. Even if the expansion of the blending ratio to 15% for model year 2001 and newer passenger vehicles were to actually occur (presently an unlikely prospect due to infrastructure limitations mentioned earlier), the higher blend wall of approximately 20 to 21 bgals would become a real barrier to expanded biofuels use by 2015.

Estimation of GHG Emission Reductions

Under EISA, EPA is responsible for evaluating whether a renewable fuel meets the specific GHG reduction threshold assigned to its RFS category. Determining compliance with the thresholds requires a comprehensive evaluation of renewable fuels on the basis of their lifecycle emissions.[93] The concept of "lifecycle emissions" encompasses an evaluation of GHG emissions along the entire pathway of a biofuel from the production, harvesting, and marketing of its feedstocks to the processing and distribution of the biofuel, including any significant indirect emissions such as emissions from land uses changes that might result from changes in crop patterns due to the various biofuels incentives (as explicitly required in Section 201, P.L. 110-140).

More specifically, some have expressed a concern that expanded field crop production in the United States for ethanol production has led to commodity price increases that, in turn, have induced increased land cultivation in other countries, and as a result, have increased net global GHG emissions.[94] The measurement of indirect land use changes (ILUC) is necessarily inexact many potential activities and countervailing forces are involved. As a result, inclusion of ILUC as part of the EPA's lifecycle GHG reduction analysis has been controversial.

Initially, EPA's lifecycle GHG reduction models proved very sensitive to assumptions regarding the extent of indirect land use changes, and suggested that some standard biofuels may not be eligible for inclusion under the RFS. EPA models were updated prior to the final RFS rule (February 2009) using newer data and produced more inclusive results. For example, corn-starch ethanol was determined to achieve a 21% reduction in GHG emissions compared to the gasoline 2005 baseline, thus just surpassing the 20% reduction threshold.[95] EPA models for estimating land use changes and other life-cycle factors involved in GHG emissions are continually re-evaluated as new or better data, methods, or analytical techniques become available. The nature of the future changes to EPA models, and their potential to include or exclude certain biofuels, remains a critical aspect of the RFS mandates and the U.S. biofuels industry's ability to meet the mandates.

Endangerment Findings for Greenhouse Gases (GHGs)

On April 2, 2007, in *Massachusetts v. EPA* (549 U.S. 497 (2007)), the U.S. Supreme Court determined that GHGs are air pollutants covered under Section 202(a) of the Clean Air Act. The Court held that EPA must determine whether or not emissions of GHGs from new motor vehicles cause or contribute to air pollution that may reasonably be anticipated to endanger public health or welfare, or whether the science is too uncertain to make a reasoned decision.[96] This court ruling allows EPA to regulate GHGs without further congressional action, and could bring into play the issue of indirect land use changes, given their alleged GHG emissions effects, which may put all ethanol production in question. On June 11, 2010, a Senate resolution (S.J.Res. 26) that would have blocked EPA from using the Clean Air Act to regulate GHGs was defeated (53-47).[97] Prior to the vote, on June 8, 2010, the White House had issued a statement saying that if S.J.Res. 26 reached the President's desk (i.e., passed both chambers of Congress), President Obama would veto it.

Other Pending or Emerging Biofuels Issues

CARB's LCFS Restriction on Midwestern Ethanol

In January 2007, then-Governor Schwarzenegger established a Low Carbon Fuels Standard (LCFS) by executive order for California.[98] The executive order directed the state's Secretary for Environmental Protection to coordinate the actions of the California Energy Commission, the California Air Resources Board (CARB), the University of California, and other agencies to develop protocols for measuring the "life-cycle carbon intensity" of transportation fuels.

Under the LCFS, CARB proposed reducing emissions of GHGs by lowering the carbon content of transportation fuels used in California. The LCFS established performance standards that fuel producers and importers must meet each year starting in 2011. Unlike the RFS, which groups biofuels into four categories, the LCFS evaluates each fuel on its own demonstrated level of lifecycle GHG emissions. The LCFS requires that biofuels demonstrate lower lifecycle GHG than the fossil fuels that they replace. For corn ethanol, carbon intensity is lowered by using natural gas instead of coal as a processing fuel, substituting biomass for natural gas or coal, and selling DDGS wet instead of dry.[99] For biodiesel and renewable diesel, carbon intensities can be lowered dramatically by using tallow or recycled cooking oils instead of soybean oil.

As part of its LCFS modeling effort, CARB includes an estimate of the indirect land use changes (ILUC) impact of grain-based ethanol. Largely because of the ILUC value assigned to corn-starch ethanol, most midwestern ethanol production did not qualify for use as a transportation fuel under California's LCFS.[100] This result has important implications for how or whether the federal RFS mandates can be met for the nation as a whole, since California is the largest state (39 million people), the largest consumer of gasoline (over 11% of national highway fuel use),[101] and a major ethanol consumer of approximately 1.5 billion gallons annually.[102]

The ILUC inclusion sparked considerable reaction from biofuel proponents because the measurement of indirect cross-country effects can be highly ambiguous.[103] In late 2010, CARB adopted a resolution to integrate the latest ILUC research into the LCFS regulation. On November 9, 2011, CARB published an updated list of CARB-approved biofuel production facilities that included 22 ethanol plants in Iowa, 21 plants in Nebraska, 12 plants in South Dakota, and 11 plants in Minnesota among the 111 newly added biofuel-plant pathways.[104] On November 26, 2012, CARB published a "Final Regulation Order" describing the LCFS compliance schedule and carbon intensity lookup table for various fuel pathways.[105]

On December 24, 2009, several ethanol groups (including RFA and Growth Energy) filed a lawsuit asserting that the California LCFS violated the U.S. Constitution by seeking to regulate farming and ethanol production practices in the United States under the "commerce clause," which leaves regulation of interstate commerce to the federal government.[106] On December 29, 2011, a U.S. district judge ruled that California's LCFS law did violate the U.S. Constitution's commerce clause and issued an injunction halting enforcement of California's LCFS. The judge ruled that CARB had failed to establish that there are no alternative methods to advance its goals of reducing GHG emissions to combat global warming. After an initial request for a stay of injunction by CARB was denied, a second request for a stay of injunction, while CARB appeals the original ruling, was filed with the Ninth District Court of Appeals and was granted as of April 23, 2012, allowing CARB to continue enforcement of the LCFS until a ruling on the appeal is made.[107]

EU Anti-Dumping Charges Issued Against U.S. Ethanol Exports

U.S. ethanol exports surged to a record 1.2 billion gallons in 2011 (Figure 7), driven in part by blending wall limits, but also motivated in part by a sharp fall-off in Brazil's ethanol exports due to high international sugar prices and a below-average sugarcane harvest. The top three destinations for U.S. ethanol exports in 2011 were Brazil (33%), Canada (25%), and the European Union (EU) (24%)—all three of which had their own national biofuels usage mandates. Large U.S. ethanol exports are problematic for two reasons—first, they run counter to the often-cited policy goal of national energy security, and second, they may conflict with biofuels policy goals in other countries, leading to trade disputes.

EU policy has promoted renewable energy use, along with GHG reductions and energy conservation, for much of the past decade.[108] As a result, EU policy support has engendered a substantial domestic renewable energy industry. As part of a "Renewable Energy Directive" adopted by the European Parliament on December 17, 2008, the EU established a 20-20-20 plan that calls for a 20% reduction in GHG emissions compared to 1990 levels, a 20% increase in renewable energy use (with a 10% share specifically in the transport sector), and a 20% reduction in overall energy consumption. As part of the 20-20-20 plan, the EU also adopted a mandate for renewable content in transportation fuels of 5.75% in 2010, rising to 10% by 2020. On October 17, 2012, the EU revised its policy proposal to state that the use of food-based biofuels to meet the 10% renewable energy target in transportation fuels of the Renewable Energy Directive will be limited to 5%.[109]

After the surge of ethanol imports from the United States in 2011, an association of European ethanol producers, ePURE, claimed that the blending tax credit—the $0.45 per gallon incentive known as VEETC—then available to U.S. biofuels blenders represented a subsidy, and that the importation of "subsidized" U.S. ethanol was hurting EU biofuel producers. As a result ePURE requested an anti-dumping (AD) and countervailing duty (CVD) investigation.

On November 25, 2011, the EU initiated an investigation into whether U.S. exporters sold ethanol at unfair prices and were backed by subsidies in violation of international trade rules to the detriment of EU biofuels producers.[110] At issue is a European allegation that international ethanol traders were exporting E90 (90% ethanol blends) to Europe to take advantage of the EU's lower tariff on such blends as well as the tax incentive for ethanol blending in the United States. In response to the EU anti-dumping investigation, the Renewable Fuels Association (RFA)[111] pointed out that the ethanol tax credits (most of which expired on December 31, 2011) were not made available to U.S. ethanol producers, but "to gasoline blenders, marketers, and other end users."[112]

After a 15-month investigation into a number of U.S. ethanol producers, the EU concluded that U.S. domestic policies aiming to encourage clean energy constitute an illegal subsidy and lead to artificially low-priced imports being "dumped" on the EU market.[113] On February 28, 2013, the European Commission announced that it will impose a five-year anti-dumping duty of 9.5% on all imports of bioethanol from the United States into the 27-nation bloc. In 2009, when similar complaints were lodged against U.S. biodiesel exports, the EU imposed duties of 40% for a five- year period on biodiesel imports originating from the United States.[114]

In response, on April 29, 2013, a bipartisan group of U.S. senators asked the U.S. Trade Representative (USTR), Demetrios Marantis, to investigate the EU decision and consider the possibility of filing a World Trade Organization (WTO) challenge to the European Commission's decision.[115]

The potential implications of an ethanol trade dispute between United States and the EU are unclear. However, the imposition of an import tariff will likely limit U.S. ethanol exports to the EU. Given the emergence of the blend wall as a constraint on U.S. ethanol consumption, combined with relatively tight ethanol supplies on the world market (following two years of successive poor Brazilian sugar crops—2011 and 2012) and biofuels usage mandates in several major fuel consuming nations, the United States may seek international markets for surplus domestic supplies, thus keeping the issue in front of policymakers.

End Notes

[1] This excludes the costs of externalities (e.g., air pollution, environmental degradation, illness and disease, or indirect land use changes and market-price effects) linked to emissions associated with burning either fossil fuels or biofuels.

[2] For more details and a complete listing of federal biofuels programs and incentives, see CRS Report R42566, *Alternative Fuel and Advanced Vehicle Technology Incentives: A Summary of Federal Programs*.

[3] See the list of related CRS Reports available at the CRS website "Issues in Focus: Agriculture: Agriculture-Based Biofuels" including CRS Report R41985, *Renewable Energy Programs and the Farm Bill: Status and Issues.*

[4] See CRS Report R40529, *Biomass: Comparison of Definitions in Legislation Through the 112th Congress.*

[5] See CRS Report R40155, *Renewable Fuel Standard (RFS): Overview and Issues.*

[6] See CRS Report R42122, *Algae's Potential as a Transportation Biofuel.*

[7] According to data from the Renewable Fuel Association, U.S. ethanol production in 2012 was 13.3 billion gallons (61%), Brazil's was 5.8 billion gallons (27%), and the world total was 21.8 billion gallons (100%).

[8] In 2012, 20,069 gallons of cellulosic biofuels production were reported to the Environmental Protection Agency (EPA) under its RFS2 EMTS Informational Data system, at http://www.epa.gov/otaq/fuels/rfsdata/. Data concerning cellulosic biofuels production costs is proprietary and has not been made publicly available.

[9] See CRS Report R41106, *Meeting the Renewable Fuel Standard (RFS) Mandate for Cellulosic Biofuels: Questions and Answers.*

[10] Energy Information Agency (EIA), *Monthly Biodiesel Production Report*, U.S. Dept. of Energy (DOE), March 2013.

[11] The RFS referred to as RFS1 was begun by the Energy Policy Act of 2005 (§1501; P.L. 109-58). A greatly expanded RFS (referred to as RFS2) was established by the Energy Independence and Security Act of 2007 (EISA, §202, P.L. 110-140). For more information on the RFS, see CRS Report R40155, *Renewable Fuel Standard (RFS): Overview and Issues*; this is described in greater detail later in this report, in the section titled "Evolution of the U.S. Ethanol Sector."

[12] Each RFS biofuel category has an identifier code associated with it: D6 is for an unspecified renewable fuel, D5 is for an advanced biofuel, D4 is for biomass-based diesel, D3 is for cellulosic biofuel, and D7 is for cellulosic diesel.

[13] CRS Report R40460, *Calculation of Lifecycle Greenhouse Gas Emissions for the Renewable Fuel Standard (RFS).*

[14] Ethanol's use as an additive for octane or oxygenate purposes occurs primarily at low blend levels of up to 5%, and is small relative to the growth in total usage of recent years. When ethanol is being added to enhance engine performance rather than as a fuel extender, it is a complement to gasoline and may potentially capture a price premium over standard gasoline.

[15] Because the RFS categories are nested, their values will include a premium to reflect a higher nesting. For example, corn ethanol only qualifies for the total renewable fuel category (D6). Ethanol from other feedstock qualifies for the more restrictive advanced biofuel category (D5) as well as the D6 category. Cellulosic ethanol also qualifies for the cellulosic biofuels category (D3) along with the D5 and D6 categories. Thus, as long as the RFS mandate is binding, a gallon of cellulosic ethanol will have inherently greater value than a gallon of advanced biofuel which itself has inherently greater value than a gallon of corn ethanol.

[16] Biodiesel qualifies for the biomass-based diesel (BBD) category (D4) which, by its nested nature, also qualifies for the advanced (D5) and total biofuel (D6) categories. If BBD is produced under a production process that uses cellulosic biomass as its originating feedstock, then it may be defined as cellulosic diesel (D7) and qualify for the nested cellulosic biofuels category (D3).

[17] RINs are discussed in more detail in CRS Report R40155, *Renewable Fuel Standard (RFS): Overview and Issues* and CRS Report R42824, *Analysis of Renewable Identification Numbers (RINs) in the Renewable Fuel Standard (RFS).*

[18] The exception is cellulosic ethanol, whose RFS mandate was waived to lower levels by EPA in each of its first four years of existence (2010-2013).

[19] Scott Irwin and Darrel Good, "Freeze It—A Proposal for Implementing RFS2 through 2015" *farmdoc-Daily*, April 10, 2013.

[20] For a discussion of the blend wall and associated policy and market issues, see CRS Report R40155, *Renewable Fuel Standard (RFS): Overview and Issues.*

[21] "Ethanol Policy: Past, Present, and Future," by James A. Duffield, Irene M. Xiarchos, and Steve A. Halbrook, *South Dakota Law Review*, Fall 2008.

[22] USDA, Office of Energy Policy and New Uses, *The Energy Balance of Corn Ethanol: An Update*, AER-813, by Hosein Shapouri, James A. Duffield, and Michael Wang, July 2002.

[23] "Status and Impact of State MTBE Ban," Energy Information Administration (EIA), U.S. Dept. of Energy (DOE), revised March 27, 3003; available at http://www.eia.doe.gov/oiaf/servicerpt/mtbeban/.

[24] The 30 million gallon threshold was extended to 60 million gallons by the Energy Policy Act of 2005 (P.L. 109-58).

[25] The Bioenergy Program was initiated on August 12, 1999, by President Clinton's Executive Order 13134. On October 31, 2000, then-Secretary of Agriculture Glickman announced that, pursuant to the executive order, $300 million of Commodity Credit Corporation (CCC) funds ($150 million in both FY2001 and FY2002) would be made available to encourage expanded production of biofuels.

[26] The CCC is a U.S. government-owned and -operated corporation, created in 1933, with broad powers to support farm income and prices and to assist in the export of U.S. agricultural products. Toward this end, the CCC finances USDA's domestic farm commodity price and income support programs and certain export programs using its permanent authority to borrow up to $30 billion at any one time from the U.S. Treasury.

[27] The Bioenergy Program was phased out at the end of FY2006.

[28] "Status and Impact of State MTBE Ban," Energy Information Administration (EIA), U.S. Dept. of Energy (DOE), revised March 27, 3003; available at http://www.eia.doe.gov/oiaf/servicerpt/mtbeban/.

[29] Ibid.

[30] Based on CRS simulations of an ethanol dry mill spreadsheet model developed by D. Tiffany and V. Eidman in *Factors Associated with Success of Fuel Ethanol Producers*, Staff Paper P03-7, Dept of Applied Economics, University of Minnesota, August 2003. Note, nameplate capacity represents the capacity that the design engineers will warrant. In most cases, an efficiently run plant will operate in excess of its nameplate capacity.

[31] See CRS Report R40155, *Renewable Fuel Standard (RFS): Overview and Issues.*

[32] For more information about markets during this period, see CRS Report RL34474, *High Agricultural Commodity Prices: What Are the Issues?* See also, "What Is Driving Food Prices," by Philip C. Abbott, Christopher Hurt, and Wallace E. Tyner, Farm Foundation, July 2008; hereinafter referred to as Abbott et al., 2008.

[33] On June 23, 2008, the nearby futures contract for No. 2, yellow corn hit a then-record $7.65 per bushel on the Chicago Board of Trade. On July 7, 2008, the nearby futures contract for Crude Oil hit $147.27 per barrel at the New York Mercantile Exchange, while the nearby Brent Crude Oil contract hit $147.50 at the ICE Futures Europe exchange.

[34] Permanent Subcommittee on Investigations, U.S. Senate, *Wall Street and the Financial Crisis: Anatomy of a Financial Collapse*, Majority and Minority Staff Report, April 13, 2011.

[35] See CRS Report R41985, *Renewable Energy Programs and the Farm Bill: Status and Issues*.

[36] See the section "2008 Farm Bill Expiration" later in this report for details.

[37] EIA, DOE, "Petroleum Products Supplied by Type;" http://www.eia.gov/totalenergy/data/monthly/pdf/sec3_15.pdf.

[38] USDA, ERS, *Outlook for U.S. Agricultural Trade*, AES-72, November 30, 2011.

[39] For more information on this and other market factors, see CRS Report R41956, *U.S. Livestock and Poultry Feed Use and Availability: Background and Emerging Issues*.

[40] And to help satisfy California's Low Carbon Fuel Standard (LCFS) described later in this report.

[41] Based on official statistics from the International Trade Commission, Dept. of Commerce.

[42] Midpoint of a projected range of $4.20 to $5.00 per bushel, *World Agricultural Supply and Demand Estimates (WASDE)*, World Agricultural Outlook Board (WAOB), USDA, June 12, 2012.

[43] *WASDE*, WAOB, USDA, August 10, 2012.

[44] Informa projects that U.S. ethanol production will fall by nearly 550 million gallons to 2013, to a level of 12.8 billion gallons—Informa Economics, "Retail Gasoline Price Impact of Compliance with the Renewable Fuel Standard," whitepaper prepared for the Renewable Fuel Association, March 25, 2013.

[45] EIA, *Monthly Energy Review*, March 2013; at http://www.eia.gov/totalenergy/data/monthly/# renewable.

[46] RINs are 38-character numeric and alpha codes generated when a qualified renewable fuel is produced or imported that move through the supply chain with the renewable blendstock and are transferred to buyers, either with physical biofuel or separated from it, as a credit. RINs are the basic currency for compliance and trades in the Renewable Fuels Standard. In RIN trade, D6 RINs for ethanol and D4 RINs for biomass diesel or biodiesel get the most attention because they are the most liquid. For information on RINs, see CRS Report R40155, *Renewable Fuel Standard (RFS): Overview and Issues* and CRS Report R42824, *Analysis of Renewable Identification Numbers (RINs) in the Renewable Fuel Standard (RFS)*.

[47] OPIS Ethanol and Biodiesel Information Service, U.S. RINs (prices in U.S. $/RIN), Ethanol & Gasoline Component Spot Market Prices, various weekly issues, January-March 2013.

[48] Scott Irwin and Darrel Good, "Exploding Ethanol RINs Prices: What's the Story?," *FarmdocDaily,* Department of Agriculture and Consumer Economics, University of Illinois, March 8, 2013, at http://farmdocdaily.illinois.edu/.

[49] See CRS Report R40155, *Renewable Fuel Standard (RFS): Overview and Issues.*

[50] See CRS Report R40155, *Renewable Fuel Standard (RFS): Overview and Issues.*

[51] See CRS Report R41106, *Meeting the Renewable Fuel Standard (RFS) Mandate for Cellulosic Biofuels: Questions and Answers.*

[52] Cellulosic biofuels are derived from the sugar contained in plant cellulose. For more information, see CRS Report R41106, *Meeting the Renewable Fuel Standard (RFS) Mandate for Cellulosic Biofuels: Questions and Answers.*

[53] EIA, *Monthly Biodiesel Production Report*, DOE, March 2013.

[54] Robert Wisner, "Feedstocks Used for U.S. Biodiesel: How Important is Corn Oil?" *AgMRC Renewable Energy & Climate Change Newsletter*, April 2013, at http://www.agmrc.org.

[55] For more information, visit the NBB at http://www.biodiesel.org.

[56] EIA, DOE; biodiesel production estimates from "Annual Energy Outlook 2013," Transportation Sector Energy Use by Mode and Type, Reference Case.

[57] See section "The Ethanol Industry's Perfect Storm in 2005."

[58] While similar to "biodiesel," "renewable diesel" is produced through different processes and results in a fuel with somewhat different chemical characteristics. There is a separate tax credit of $1.00 per gallon for renewable diesel.

[59] DOE, EIA, *Monthly Biodiesel Production Report*, March 2009.

[60] EPA, "EPA Proposes 2013 Renewable Fuel Standards," EPA-420-F-13-007, January 2013.

[61] "EU Imposes Five-Year AD, CVD Duties on U.S. Biodiesel," *Inside U.S. Trade*, July 7, 2009.

[62] EIA, *Monthly Energy Review*, March 2013, Table 10.4 "Biodiesel Overview."

[63] "Tax Credits, Mandates Bring Back Biodiesel Plants," *Energy & Environmental New*, September 19, 2011.

[64] Robert Wisner, "Feedstocks Used for U.S. Biodiesel: How Important is Corn Oil?" *AgMRC Renewable Energy & Climate Change Newsletter*, April 2013; at http://www.agmrc.org.

[65] Federal Highway Administration, U.S. Deptartment of Transportation, "State Motor-Vehicle Registration—2011," March 2013, at http://www.fhwa.dot.gov/policyinformation/statistics/ 2011/pdf/mv1.pdf.

[66] Renewable Fuel Association, "E85," at http://www.ethanolrfa.org/pages/e-85.

[67] CRS Report R40445, *Intermediate-Level Blends of Ethanol in Gasoline, and the Ethanol "Blend Wall"*.

[68] Data is from EIA/DOE's 2013 Annual Energy Outlook. EIA also projects the U.S. national biodiesel transportation fuel market to show slow but steady growth (at about 1% per year) from about 47 bgals in 2012 to nearly 54 bgals by 2022. As a result, RFS BBD's share of the biodiesel transportation fuel market is projected to remain steady at about 2.5% through 2022.

[69] EPA, Fuels and Fuel Additives, "EPA Announces E15 Partial Waiver Decision and Fuel Pump Labeling Proposal," EPA420-F-10-054, October 13, 2010; at http://www.epa.gov/otaq/regs/fuels/additive/e15/420f10054.htm.

[70] See EPA, "E15 (a blend of gasoline and ethanol)," at http://www.epa.gov/otaq/regs/ fuels/additive/e15/.

[71] For more information on potential misfueling, see CRS Report R40155, *Renewable Fuel Standard (RFS): Overview and Issues*

[72] "E15 Decision Opens Blend to 2 Out of 3 Vehicles; More Work Yet to be Done," RFA news release, Jan. 21, 2011.

[73] For more information, see the Renewable Fuels Association's E-85 online information site at http://www.ethanolrfa.org/pages/e-85.

[74] Based on CRS calculations using EIA and USDA data.

[75] Examples of ethanol policy proponents include the Renewable Fuels Association (RFA), the National Corn Growers Association (NCGA), and Growth Energy. Biodiesel proponents include the American Soybean Association and the National Biodiesel Board.

[76] Advocates of this position include free-market proponents such as the Cato Institute, federal budget watchdog groups such as Citizens Against Government Waste, Taxpayers for Common Sense, and farm subsidy watchdog groups such as the Environmental Working Group.

[77] For example, see James and Stephen Eaves, "Is Ethanol the 'Energy Security' Solution?" editorial, Washingtonpost.com, October 3, 2007; or R. Wisner and P. Baumel, "Ethanol, Exports, and Livestock: Will There be Enough Corn to Supply Future Needs?," *Feedstuffs*, no. 30, vol. 76, July 26, 2004.

[78] For example, the American Petroleum Institute (API) and the American Fuel & Petrochemical Manufacturers (AFPM) have brought legal challenges against certain aspects of federal biofuels programs.

[79] For a more complete list of federal biofuels incentives, see CRS Report R40110, *Biofuels Incentives: A Summary of Federal Programs*.

[80] See CRS Report RS21930, *Ethanol Imports and the Caribbean Basin Initiative (CBI)*.

[81] RFS (referred to as RFS1) was begun by the Energy Policy Act of 2005, (§ 1501; P.L. 109-58). The RFS was greatly expanded (referred to as RFS2) by the Energy Independence and Security Act of 2007 (EISA, § 202, P.L. 110-140). For more information on the RFS, see CRS Report R40155, *Renewable Fuel Standard (RFS): Overview and Issues*.

[82] See the earlier section, "The Renewable Fuel Standard (RFS)," for more details.

[83] CRS Report R40460, *Calculation of Lifecycle Greenhouse Gas Emissions for the Renewable Fuel Standard (RFS)*.

[84] CRS Report R41985, *Renewable Energy Programs and the Farm Bill: Status and Issues*.

[85] Originally the allowance was for cellulosic biofuel plant property. However, P.L. 112-240 amended the credit to included plant property used for non-cellulosic fuel produced from algae feedstocks. The special depreciation allowance involves 50% of the adjusted basis of a new cellulosic or algae-based biofuel plant in the year it is put in service, less any portion of the cost financed via tax-exempt bonds.

[86] For more information, see the "Federal & State Incentives & Laws," Alternative Fuels and Advanced Vehicles Data Center, Energy Efficiency and Renewable Energy (EERE), DOE, at http://www.afdc.energy.gov/afdc/laws/.

[87] For more information on U.S. farm programs, see CRS Report RL34594, *Farm Commodity Programs in the 2008 Farm Bill*; CRS Report R40422, *A 2008 Farm Bill Program Option: Average Crop Revenue Election (ACRE)*; CRS Report R40452, *A Whole-Farm Crop Disaster Program: Supplemental Revenue Assistance Payments (SURE)*; and CRS Report R40532, *Federal Crop Insurance: Background.*

[88] Mandatory funding is derived from authorizing legislation and is not subject to annual appropriations.

[89] For details see CRS Report R42442, *Expiration and Extension of the 2008 Farm Bill.*

[90] See CRS Report R41985, *Renewable Energy Programs and the Farm Bill: Status and Issues.*

[91] See CRS Report R41296, *Biomass Crop Assistance Program (BCAP): Status and Issues.*

[92] See CRS Report RS22870, *Waiver Authority Under the Renewable Fuel Standard (RFS).*

[93] For more information, see CRS Report R40460, *Calculation of Lifecycle Greenhouse Gas Emissions for the Renewable Fuel Standard (RFS).*

[94] Tim Searchinger et al., "Use of U.S. Croplands for Biofuels Increases Greenhouse Gases Through Emissions from Land-Use Change," *Science*, Vol. 319 no. 5867, February 29, 2008, pp. 1238-1240.

[95] "V. Lifecycle Analysis of Greenhouse Gas Emissions;" Regulation of Fuels and Fuel Additives: Changes to Renewable Fuel Standard Program; Final Rule, 40 CFR Part 80, *Federal Register*, March 26, 2010, p. 14786.

[96] For more information see "Endangerment and Cause or Contribute Findings for Greenhouse Gases under Section 202(a) of the Clean Air Act," EPA, at http://www.epa.gov/climatechange/endangerment.html.

[97] DTN Ag Policy Blog, "Senators Face Emissions Test," Chris Clayton, June 9, 2010.

[98] For more information, see "Low Carbon Fuel Standard," California Energy Commission, at http://www.energy.ca.gov/low_carbon_fuel_standard/index.html.

[99] EIA, "Biofuels Issues and Trends," October 2012, p. 25, at http://www.eia.gov.

[100] For more information see, "Proposed Regulation to Implement the Low Carbon Fuel Standard," Initial Statement of Reasons, Vol. 1, CARB, March 5, 2009, at http://www.arb.ca.gov/fuels/lcfs/lcfs.htm.

[101] Federal Highway Administration, Dept. of Transportation, "U.S. Motor Fule Use- 2011," Table MF-21, February 2013; at http://www.fhwa.dot.gov/policyinformation/statistics/2011/mf21.cfm.

[102] Todd Neeley, "US Scientists Demand Revision of Biofuels Carbon Accounting," DTN Ethanol blog, May 25, 2010.

[103] A 2010 analysis from Purdue University concluded that CARB overestimated the ILUC impact of grain-based ethanol by a factor of two in developing its LCFS: "New Study Undercuts California Low Carbon Fuel Standard, Shows Evolving Land Use Change Debate," Renewable Fuels Association (RFA), news entry, April 28, 2010, at http://www.ethanolrfa.org. The final version of the Purdue study was released as Wallace E. Tyner, Farzad Taheripour, Qianlai Zhuang, Dileep Birur, and Uris Baldos, *Land Use Changes and Consequent CO2 Emissions due to US Corn Ethanol Production: A Comprehensive Analysis*," Department of Agricultural Economics, Purdue University, July 2010, at http://www.transportation.anl.gov/pdfs/MC/625.PDF. Researchers at DOE's Oak Ridge National Laboratory (ORNL) concluded that ILUC resulting from expanded corn ethanol production over the past decade has likely been minimal to zero: RFA, "Dept. of Energy Researchers: ILUC Impact 'Minimal to Zero,'" 2010 press releases, October 20, 2010, at http://www.ethanolrfa.org; and Debo Oladosu and Keith Kline, ORNL, "Empirical Analysis of the Sources of Corn Used for Ethanol Production in the United States: 2001-2009," presentation to National Corn Growers Association, November 4, 2010; at http://www.ornl.gov/sci/besd/cbes/Symposia/Empirical_Analysis_Source_Corn_Ethanol_Nov2010.pdf.

[104] CARB, "Fuel Production Facilities with ARB Approved Physical Pathway Demonstrations," LCFS Program, November 9, 2012, at http://www.arb.ca.gov/fuels/lcfs/lcfs.htm.

[105] CARB, "Final Regulation Order (unofficial electronic version)," November 26, 2012, at http://www.arb.ca.gov/ fuels/lcfs/lcfs.htm.

[106] Todd Neely, "Court Strikes Down California LCFS: Ruling Opens Door to Large Ethanol Market," *DTN: The Progressive Farmer*, December 29, 2011.

[107] EIA, "Biofuels Issues and Trends," DOE, October 2012.

[108] European Commission, "Renewable Energy;" at http://ec.europa.eu/energy/renewables/.

[109] European Commission, "Renewable Energy/Targets by 2020;" at http://ec.europa.eu/energy/ renewables/targets_en.htm.

[110] *World Trade Online*, "EU Launches AD, CVD Investigations into U.S. Ethanol Exports," December 8, 2011.

[111] The RFA is a U.S. ethanol

[112] "RFA Responds to EU Ethanol Investigation," RFA News Release, November 28, 2011; http://www.ethanolrfa.org.

[113] International Centre for Trade and Sustainable Development (ICTSD), "Disputes Roundup: Trade Remedies in the Spotlight in Geneva, Brussels," Bridges Weekly Trade News Digest, Vol. 17, No. 7, February 27, 2013.

[114] *Agri-Pulse*, "RFA Tells EU: 'It's Not Us!" November 30, 2011.

[115] Todd Neely, "Anti-Dumping Questioned," *DTN Progressive Farmer*, April 30, 2013.

In: Agriculture-Based Biofuels ...
Editor: Doug R. Monty

ISBN: 978-1-63117-783-5

Chapter 2

THE FEASIBILITY OF PRODUCING AND USING BIOMASS-BASED DIESEL AND JET FUEL IN THE UNITED STATES*

A. Milbrandt, C. Kinchin and R. McCormick

ABSTRACT

This study summarizes the best available public data on the production, capacity, cost, market demand, and feedstock availability for the production of biomass-based diesel and jet fuel. It includes an overview of the current conversion processes and current state-of-development for the production of biomass-based jet and diesel fuel, as well as the key companies pursuing this effort. The discussion analyzes all this information in the context of meeting the Renewable Fuels Standard mandate, highlights uncertainties for the future industry development, and key business opportunities.

ABBREVIATIONS, ACRONYMS, AND DEFINITIONS

ATJ	alcohol-to-jet fuel
ASTM	American Society for Testing and Materials

* This is an edited, reformatted and augmented version of Technical Report NREL/TP-6A20-58015, issued by the National Renewable Energy Laboratory, December 2013.

CAFE	Corporate Average Fuel Economy
C&D	construction and demolition
DoD	Department of Defense
EIA	Energy Information Administration
EISA	Energy Independence and Security Act
EPA	Environmental Protection Agency
EtOH	Ethanol
FAME	fatty acid methyl esters
FOG	fats, oils, and greases
FT	Fischer Tropsch
gge	gasoline gallon equivalent
GHG	greenhouse gas
ha	hectare
HEFA	hydrogenated esters and fatty acids
IPK	iso-paraffinic kerosene
LPG	liquefied petroleum gas
mpg	miles per gallon
MSW	Municipal solid waste
PADD	Petroleum Administration for Defense District
R&D	research and development
RFS	Renewable Fuels Standard
TAG	triacylglycerol
TEA	techno-economic analysis

Executive Summary

In recent years, growing concern about U.S. dependence on oil imports and greenhouse gas (GHG) emissions has resulted in greater interest in alternative fuels that can be produced from domestic renewable feedstock. Currently, diesel and jet fuel substitutes derived from biomass material—such as crop and forest residues, dedicated energy crops, and plant oils—are receiving increasing attention.

The purpose of this study is to evaluate the technical and economic feasibility of producing and using biomass-based diesel and jet fuel in the United States. To achieve this goal, data on production, capacity, cost, market demand, and feedstock availability were gathered and analyzed. Environmental considerations and legislative climate were not evaluated here, but these factors would contribute to an important follow-up study and provide a more complete picture of the biomass-based diesel and jet fuel potential in the United States.

Some of the key findings of this study include the following:

1) It is technically feasible to produce biomass-derived diesel and jet fuel substitutes in the United States. Many conversion technology options exist. Some are commercially available or in demonstrational stage; others are still in the research and development phase.
2) Biodiesel, consisting of fatty acid methyl esters (FAME) produced from lipids (fats, oils, and greases), is currently the predominant form of biomass-based diesel. Production reached a record 1.1 billion gallons in 2011 and kept at that level in 2012. It is expected to be higher in 2013. Biodiesel blends cannot yet be considered fully "drop-in" fuels because they cannot be transported in all petroleum product pipelines. For pipelines that transport jet fuel, there is a concern that the jet fuel will be contaminated with biodiesel, making it unsuitable for use. Ongoing research aims to determine what, if any, level of FAME can be tolerated in jet fuel.
3) In comparison, the current U.S. renewable diesel and jet fuel production capacity is small, about 225 million gallons per year. These fuels can be produced from various biomass resources and through several different approaches which all target hydrocarbon products that are similar to petroleum fuels in chemical makeup, and therefore may be considered "drop-in" fuels. It is anticipated that, as "drop-in" fuels, they can be blended with petroleum diesel/jet fuel at high levels, or possibly used in neat form.
   - Renewable diesel is produced at commercial scale primarily by hydroisomerization of lipid feedstock. Currently, there are two commercial facilities utilizing this process: Dynamic Fuels, a joint venture between Syntroleum Corporation and Tyson Foods (Geismar, Louisiana) and Diamond Green Diesel, a joint venture between Valero subsidiary Diamond Alternative Energy LLC and Darling International Inc. (Norco, Louisiana).
   - A number of processes are under development for production of renewable diesel from biomass-derived sugars (corn, sugarcane, and sorghum, as well as sugars from thermochemical or biochemical depolymerization of cellulose and hemicellulose).
   - Processes are also being developed for direct conversion of lignocellulosic biomass by fast pyrolysis, gasification, and other

thermochemical means. The first commercial plant is operated by KiOR in Columbus, Mississippi and became operational in early 2013.

4) The costs for producing renewable diesel and jet fuel are not well known and involve a high degree of uncertainty. The process economics for these fuels is highly dependent upon the cost of the feedstock, similar to biodiesel. Additionally, variables such as plant size and co-product credits can have a significant impact on the overall production cost. Hydroisomerization of lipids is performed commercially by Dynamic Fuels and Diamond Green Diesel and internationally by Neste Oil. The KiOR technology, utilizing pyrolysis, is at an initial commercial scale. Other technology routes are not yet commercial and display a wide range of estimated costs in public sources. As these technology pathways mature and become more widespread, more specific information regarding their economics will be available, which will enable a more detailed analysis and performance comparison.
5) From a feedstock perspective, enough lignocellulosic material is projected to be available in support of the Renewable Fuels Standard (RFS) mandate of 21 billion gallons of advanced biofuels. Crop and forest residues alone could yield about 8-24 billion gallons of biomass-based diesel/jet fuel in 2022 (assuming a conversion via fast pyrolysis).This potential could be larger if the conversion technologies achieve higher yields and if additional feedstock, such as dedicated energy crops, become available. However, there will be competition for lignocellulosic feedstock with the ethanol industry and renewable gasoline producers to meet the RFS mandate. Thus, it is unclear what share the renewable diesel/jet fuel would have in the total biofuels contribution. Ultimately, it will depend on the rate of commercialization of these technologies, selling price, and the transportation market demands.
6) Based on current statistics, and proven by the biodiesel industry, there is enough lipid feedstock to support the production of 1 billion gallons of biomass-based diesel mandated by the RFS. Today, roughly half of the biodiesel in the United States is produced from soybeans. The remaining portion consists of animal fat, used cooking oil, canola, and some other minor feedstocks. While soybean production is projected to grow in coming years, the biodiesel industry hopes to achieve higher output through advanced technologies for increasing oil supply

and production of new feedstock. If algal oil becomes commercially available, as projected within the next 5-10 years, it would greatly benefit both biodiesel and renewable diesel/jet fuel industries. Given the right resources, algal oil productivity can be quite high. Algae are a potential aquatic oil crop, but may also yield carbohydrates that can be converted to sugar.

7) Demand for diesel and jet fuel in the United States is projected to grow. As easily recoverable crude oil resources are diminishing and as their prices rise, more substitutes are expected to enter the market.
   - Among the diesel consumers in the country, freight trucking has the largest share. The number of light-duty vehicles using diesel is projected to increase, the rate of which will depend on the market penetration of other alternatives such as hybrid and electric vehicles.
   - Jet fuel is forming as a large and profitable market for the renewable fuels industry. Success in this area could stimulate a significant increase in the production of biofuels and associated feedstock. It is expected that jet fuel consumption by commercial carriers will continue to grow over the next years, whereas jet fuel consumption by the military will remain flat.
8) For biomass-based diesel and jet fuel to be successful among the trucking and aviation companies, they must be cost-competitive with petroleum-based fuels. It is uncertain what the future holds for these substitutes, but it is expected that the next several years, as more facilities come online, will answer many questions about the economic viability of these technologies. Much will depend on the rate of recovery of U.S. and world economies, oil prices, carbon market, and political climate.

INTRODUCTION

In recent years, growing concern about U.S. dependence on oil imports and greenhouse gas (GHG) emissions has resulted in greater interest in alternative fuels that can be produced from domestic renewable feedstock. Currently, diesel and jet fuel substitutes derived from biomass material—such as crop and forest residues, dedicated energy crops, and plant oils—are receiving increasing attention.

The purpose of this study is to evaluate the technical and economic feasibility of producing and using biomass-based diesel and jet fuel in the United States. To achieve this goal, data on production, capacity, cost, market demand, and feedstock availability were gathered and analyzed. Environmental considerations and legislative climate were not evaluated here, but these factors would contribute to an important follow-up task to provide a more complete picture of the biomass-based diesel and jet fuel potential in the United States.

Overview

The Renewable Fuel Standard (RFS), a program introduced by the Energy Independence and Security Act of 2007 (EISA), defines biomass-based diesel as follows: (a) renewable fuel made from biomass; (b) meeting the definition of either biodiesel (mono-alkyl esters) or non-ester renewable diesel; (c) with lifecycle GHG emissions at least 50% less than the diesel fuel it displaces; and (d) which excludes renewable fuel derived from co-processing biomass with a petroleum feedstock[1] (EPA 2009). The jet fuel substitutes fall under the "additional renewable fuel" category of RFS2, defined as fuel produced from renewable biomass that is used to replace or reduce fossil fuels used in home heating oil or jet fuel (EPA 2009).

There are several technologies and processes that produce biomass-based diesel and jet fuel. Some of these technologies are in commercial or pre-commercial production while others are still in the research and development phase. The different technologies use a variety of feedstocks, including lignocellulosic biomass (such as wood and crop residues, dedicated herbaceous or tree energy crops), grains, sugar crops, vegetable oil (soybean, canola/rapeseed, etc.), animal fat (beef tallow, pork lard), and waste cooking greases. Table 1 summarizes the different processes used to create biomass-based diesel and jet fuel, as well as examples of companies involved in these technologies.

Biodiesel is currently the predominant form of biomass-based diesel. It is a liquid fuel made from recycled or virgin vegetable oils and animal fats through a chemical process (transesterification, described below) to produce chemical compounds known as fatty acid methyl esters (FAME). Biodiesel is the name given to these esters when they meet specifications such as ASTM D6751 or European Norm EN14214 for use as transportation fuel. Biodiesel is used in its pure form or in blends with petroleum diesel. Blends containing up to 5% volume are considered the same as conventional diesel and are fully compatible with all engines and infrastructure. Blends containing 6-20% volume biodiesel are accepted by many engine manufactures and are compatible with underground storage tanks. Fuel dispensers and related equipment that are compatible with B6 to B20 blends are available. Biodiesel blends cannot yet be considered fully "drop-in" fuels because they cannot be transported in all petroleum product pipelines. For pipelines that transport jet fuel, there is a concern that the jet fuel will be contaminated with biodiesel, making it unsuitable for use. Ongoing research aims to determine what, if any, level of FAME can be tolerated in jet fuel, and in the meantime, biodiesel is being shipped on pipelines that do not transport jet fuel. A few engine manufacturers approve of the use of B100 (pure biodiesel) in some engine models.

Renewable diesel refers to diesel fuel substitutes derived from biomass sources that chemically are not esters and thus are distinct from biodiesel. Renewable diesel and jet fuel can be produced by several different approaches (Table 1), which all target hydrocarbon products that are similar to petroleum diesel/jet fuel in chemical makeup and therefore may be considered "drop-in" fuels. It is anticipated that, as "drop-in" fuels, they can be blended with petroleum diesel/jet fuel at high levels or possibly be used in neat form. They are also expected to be transportable by pipeline.

Biodiesel from Transesterification

The process of transesterification is used for the conversion of triglycerides (the main component of vegetable oils and animal fats) to biodiesel. In this process, the feedstock is chemically reacted with an alcohol (usually methanol) in the presence of a catalyst like lye. The products are glycerin and the biodiesel fuel or FAME. These separate into a bottom glycerin layer and a top FAME layer, which can then be physically segregated.

Renewable Diesel and Jet Fuel from Hydroprocessing of Lipid Feedstocks

Renewable diesel, sometimes referred to as "green diesel" or hydrogenated esters and fatty acids (HEFA), can be produced from fatty acids (fats, oils, and greases [or FOG]) by the traditional hydroprocessing and hydroisomerization technology used in petroleum refineries. Hydroprocessing is the process of reacting feedstock with hydrogen under elevated temperatures and pressures and in the presence of a catalyst in order to remove oxygen, sulfur, and nitrogen and saturate double bonds.

Table 1. Conversion Processes for Biomass-based Diesel and Jet Fuel

Technology	Process	Fuel Produced	Companies
Lipids (fats, oils, and greases)			
	Transesterification	Biodiesel	Imperium Renewables, Renewable Energy Group, ADM, Amerigreen Energy, Inc., Cargill Inc., Direct Fuels, etc.
	Transesterification of microalgae	Biodiesel	Cellana, Solix, Seambiotic, LiveFuels
	Hydroprocessing	Diesel, Jet Fuel	Neste Oil, Dynamic Fuels LLC, Diamond Green Diesel LLC, UOP, AltAir, Emerald Biofuels LLC, etc.
	Hydroprocessing of microalgae	Diesel, Jet Fuel	Sapphare, Solazyme
Lignocellulosic Biomass/Sugars from Cellulose			
Biochemical	Conversion of cellulose via carboxylic acid	Diesel, Jet Fuel	Terrabon
	Synthetic Biology	Diesel, Jet Fuel	Amyris, LS9, Joule
Thermochemical	Gasification/Fischer Tropsch	Diesel, Jet Fuel	Choren, Flambeau River Biofuels, ClearFuels/ Rentech, TRI, Syntroleum
Hybrid Biochemical/ Thermochemical	Syngas fermentation	Diesel, Jet Fuel	Coskata, Lanztech, INEOS Bio
	Acetic acid production and lignin gasification	Diesel, Jet Fuel	ZeaChem
Depolymerization	Catalytic depolymerization of cellulose	Diesel, Jet Fuel	Covanta, Green Power
	Pyrolysis	Diesel, Jet Fuel	Envergent (UOP/Ensyn), Dynamotive, KiOR, RTI, GTI
	Thermal depolymerization	Diesel, Jet Fuel	Changing World Technologies Inc.
Other	Catalytic reforming of sugars from cellulose	Diesel, Jet Fuel	Virent
	Alchohol-to-jet fuel	Jet Fuel	Gevo, Cobalt

Note: The list of companies in the table is shown as an example and it is not meant to be complete. Some of the listed companies (such as Terrabon and Choren) were inactive at the time of writing. However, these companies are associated with either pioneering or further developing a process, and thus, they are included in this summary for illustrative purposes.

For processing of triglycerides, some of the oxygen can also be removed as carbon dioxide by decarboxylation in some processes (Kalnes et al., 2007). While the feedstock can be hydroprocessed as a co-feed with petroleum, in order to qualify as a renewable fuel (according to the RFS) the diesel must be produced as a dedicated feed in a stand-alone process. The products consist predominantly of isoparaffins with some residual normal paraffins (Smagala et al., 2013). The degree of isomerization can be adjusted to lower the cloud point, and it is even possible to obtain cloud point in the jet fuel range (less than -40°F). However, increased isomerization reduces the cetane number and leads to increased production of naphtha as a by-product. Nevertheless, for fuels with cloud points in the diesel fuel range, the cetane number will likely be over 80.

Renewable Diesel and Jet Fuel from Cellulose via Carboxylic Acid

Terrabon, a company involved in creating renewable fuels using carboxylic acid, describes the process as follows:

> It begins by treating the feedstock with lime to enhance its digestibility, and then fermenting the biomass using a mixed-culture of microorganisms to produce a mixture of carboxylic acids. Calcium carbonate is added to the fermentation to neutralize the acids to form corresponding carboxylate salts, which are then dewatered, concentrated, dried and thermally converted to ketones. The ketones are then hydrogenated to alcohols that can be refined into renewable gasoline, diesel or jet fuel blendstocks (*Ethanol Producer Magazine* 2011, 3).

Renewable Diesel and Jet Fuel Using Synthetic Biology

Synthetic biology modifies existing biological systems or builds new systems to produce novel substances. Amyris (2012) has developed genetic engineering and screening technologies that enable modification of the way microbes process (i.e., metabolize) sugar. By controlling these metabolic pathways, Amyris is able to design microbes, primarily yeast, to be living factories that convert plant-sourced sugars from crops such as sugarcane or sweet sorghum into target molecules. Using its industrial synthetic biology platform, Amyris develops yeast strains designed to produce a broad range of molecules. Amyris's building block molecule is Biofene (an Amyris-brand farnesene), a hydrocarbon molecule that can replace petrochemicals in a wide variety of products including transportation fuels such as diesel and jet fuel.

LS9 has focused on developing renewable petroleum products using a one-step fermentation process. In July 2010, the company announced the discovery of novel genes that, when expressed in *E.coli*, produce alkanes—the primary hydrocarbon components of gasoline, diesel, and jet fuel. This discovery is the first description of the genes responsible for alkane biosynthesis and the first example of a single-step conversion of sugar-to-fuel-grade alkanes by an engineered microorganism. A spokesperson for the company describes the process as a one-step sugar-to-diesel process that does not require elevated temperatures, high pressures, toxic inorganic catalysts, hydrogen, or complex unit operations (Greentech Media 2010). Similarly, the Joule process uses optimized microorganisms that act as living catalysts to produce fuel rather than first producing biomass and later extracting lipids or sugars for a subsequent multistep conversion into fuel (European Biofuels Technology Platform [EBTP] 2012).

Renewable Diesel and Jet Fuel from Biomass Gasification (Fischer-Tropsch)

Another process for making renewable diesel and jet fuel is converting cellulosic biomass through high-temperature gasification into synthetic gas or "syngas", a gaseous mixture rich in hydrogen and carbon monoxide. Next, a Fischer-Tropsch (FT) process is used to catalytically convert the syngas to liquid and wax products that can be refined into synthetic fuels. The production of FT liquids is a commercial technology applied to coal and natural gas.

Renewable Diesel and Jet Fuel from Syngas Fermentation

This hybrid technology is currently used to produce ethanol. However, the companies involved in this process are working on diesel/jet fuel production as well. The strategy involves the gasification of biomass to syngas before processing it into ethanol using a biochemical fermenter. Coskata (2011) describes the process: during gasification, the biomass material is converted into syngas using well-established gasification technologies. After the chemical bonds are broken in the process, Coskata's proprietary microorganisms convert the resulting syngas into ethanol by consuming the carbon monoxide (CO) and hydrogen (H2) in the gas stream. Once the gas-to-liquid conversion process has occurred, the resulting ethanol is recovered from the solution using proven distillation methods.

Renewable Diesel and Jet Fuel from Acetic Acid Production and Lignin Gasification

Similar to the hybrid process above, this technology is currently used to produce ethanol, but the company involved in this process, ZeaChem Inc., has developed a platform capable of producing other fuels as well. After fractionating the biomass, the sugar stream (both xylose [C5] and glucose [C6]) are sent to fermentation where an acetogenic process is utilized to ferment the sugars to acetic acid without CO_2 as a by-product. In comparison, traditional yeast fermentation creates one molecule of CO_2 for every molecule of ethanol. Thus, the carbon efficiency of the ZeaChem fermentation process is nearly 100% vs. 67% for yeast.

The acetic acid is converted to an ester, which can then be reacted with hydrogen to make ethanol. To get the hydrogen necessary to convert the ester to ethanol, ZeaChem takes the lignin residue from the fractionation process and gasifies it to create a hydrogen-rich syngas stream. The hydrogen is separated from the syngas and used for ester hydrogenation, and the remainder of the syngas is burned to create steam and power for the process. The net effect of combining the two processes is that about two-thirds of the energy in the ethanol comes from the sugar stream and one-third comes from the lignin steam in the form of hydrogen (ZeaChem 2011).

Renewable Diesel and Jet Fuel from Catalytic Depolymerization of Cellulose

The catalytic depolymerization process uses heat and catalysts to break long chain polymers of hydrogen, oxygen, and carbon into short-chain petroleum hydrocarbons.

> The Green Power process catalytically depolymerizes cellulosic feedstocks at moderate temperatures into liquid hydrocarbon fuels. The feedstock is first ground to a size finer than 5 mm and then placed--along with a catalyst, a low loading of lime that serves as a neutralizing agent, and a fuel that provides a liquid medium--into a reactor and heated to around 662°F. In the reactor, the feedstock is catalytically converted to liquid fuels which primarily fall within the gasoline and diesel fuel boiling ranges, although these fuels may need further upgrading. The liquid fuels are separated from any solids which are present and are

distilled into typical fuel streams including naphtha, diesel fuel, kerosene, and fuel oil (EPA 2010a, p. 42261).

Renewable Diesel and Jet Fuel from Pyrolysis Oil

Another technique for producing renewable diesel uses pyrolysis, the chemical decomposition of organic materials at elevated temperatures in the absence of oxygen. During this process, large polymers (i.e., cellulose, hemicellulose, lignin, and proteins of organic waste streams) are converted into smaller molecules and produce organic vapors, gases, and a solid residue containing carbon and ash. The vapors are condensed to produce pyrolysis oil (often referred to as bio-oil) that is then refined into diesel-like fuel. Yields of raw bio-oil as high as 75% of the initial dry weight of the biomass can be achieved (PNNL 2009).

Renewable Diesel and Jet Fuel from Thermal Depolymerization

Thermal depolymerization is similar to the geological processes thought to be involved in the production of fossil fuels—except that the technological process occurs in a timeframe measured in hours. Biomass is reacted in water at elevated temperature and pressure to form oils, gases, carbons, and ash. Hydrothermal conversion temperatures are typically 570°-660°F, with pressure sufficient to keep the water primarily as liquid (100-170 atm). The resulting bio-oil could be upgraded to a hydrocarbon product consistent with gasoline and diesel. The technology is being commercialized in the United States by Changing World Technologies (CWT). The National Advanced Biofuels Consortium (NABC) is working on further developing this process. NABC's researchers have achieved bio-oil yields of about 50% on a carbon basis from two feedstocks— wood residues and corn stover (NABC 2012b).

Renewable Diesel and Jet Fuel from Catalytic Reforming of Sugars from Cellulose

Catalytic upgrading of sugars to hydrocarbons involves separating sugars from biomass (e.g., milled corn stover) through a series of chemical and biochemical processes and catalytically upgrading it into hydrocarbon fuels.

The process has been researched by Virent, Inc. Virent's BioForming process integrates the company's patented aqueous phase reforming (APR) technology with conventional catalytic processes. First, the lignocellulosic biomass is pretreated, followed by enzymatic hydrolysis (saccharification) of the remaining cellulose. Next is catalytic conversion of the resulting glucose, xylose, and other solubilized carbon components to hydrocarbon fuels in the gasoline, jet, and diesel fuel ranges (Biddy and Jones 2013).

Alcohol-to-Jet Fuel

Alcohol-to-jet fuel (ATJ) converts short carbon chain alcohols (such as methanol, ethanol, and butanol) to the longer C12/C16 alkanes of jet kerosene. The alcohol is produced conventionally (sugar/starch fermentation), thermochemically (e.g., gasification with upgrading), or through other pathways (industrial microbiology and algae). Several companies are exploring ATJ pathways, including Gevo and Cobalt. Gevo has developed a proprietary integrated fermentation technology (GIFT) consisting of a yeast biocatalyst that converts sugars into isobutanol. The alcohol is then converted into iso-paraffinic kerosene (IPK), a blendstock used in jet fuel, via additional reactions such as dehydration, oligomerization, hydrogenation, and distillation (Gevo 2011). Similarly, Cobalt has developed its own process for extracting sugars from biomass and converting them directly into bio n-butanol, a platform molecule for the production of a broad array of fuels and chemicals, including jet fuel (Cobalt 2013).

Capacity and Production

Current U.S. biodiesel production capacity is more than 1.8 billion gallons, with about 160 plants registered with the Environmental Protection Agency (EPA) under the RFS program (*Soybean Review* 2011). The industry struggled in 2010 due to increased feedstock prices and the expiration of a key $1.00/gallon blender tax credit at the beginning of the year. According to the Energy Information Administration (EIA), biodiesel production decreased from 678 million gallons (Mgal) in 2008 to 343 Mgal in 2010 (EIA 2012a). The revised RFS (RFS2, initiated in July 2010) mandates the blending of 1 billion gallons per year of biomass-based diesel; biodiesel produced from soybean oil, animal fat, waste grease, and several other feedstocks qualifies for meeting this mandate. Driven by the RFS2 mandate and reinstatement of the $1/gallon blender tax credit, biodiesel production reached a record 1.1 billion gallons in 2011, kept that level in 2012, and it is expected to be higher in 2013 (EPA 2013a).

In comparison, the current U.S. renewable diesel and jet fuel production capacity is small, about 225 million gallons per year (Mgy) (*Biofuels Digest* 2012, KiOR 2013). The industry is just starting out and most of the facilities are at pilot/demonstration scale, under construction, or in planning phase. At the time of writing, there were three commercial facilities as described below.

Dynamic Fuels, a 50/50 joint venture between Syntroleum Corporation and Tyson Foods is located in Geismar, Louisiana. The facility has a capacity of 75 Mgy and became operational in November 2010. The plant uses animal fats, greases, and vegetable oils as feedstock. In the summer of 2011, Syntroleum announced that Dynamic Fuels achieved a record production of renewable fuels, about 87% of plant's capacity (Syntroleum 2011). High production level (71% of plant's capacity) was kept in 2012 as well (*The City Wire* 2013a).

However, the plant was idled in November 2012 because of deteriorating market conditions and the partners opted for replacement of a catalyst in the facility that would increase production efficiency. That catalyst was installed in June 2013 but the plant still remains idle today, mainly because the partners can't reach amicable restart terms (*The City Wire* 2013b).

KiOR's production facility in Columbus, Mississippi became operational in early 2013. It uses woody biomass as feedstock and has 13 Mgy capacity (KiOR 2013). Given that the facility produces both renewable gasoline and diesel, it is unknown at this time the share of renewable diesel in the final output.

Diamond Green Diesel, a joint venture between Valero subsidiary Diamond Alternative Energy LLC and Darling International Inc. is located in Norco, Louisiana. The facility has a capacity of 137 Mgy and became operational in the summer of 2013. The plant uses recycled animal fat and used cooking oil as feedstock.

It is projected that other companies will come online within the next several years, although some may produce renewable gasoline instead of, or in addition to, renewable diesel and jet fuel (*Biofuels Digest* 2012).

PRODUCTION COST

This section presents production cost estimates for several technology pathways. It is important to note that the costs presented here are production costs rather than final selling prices at the pump.

Very little data are available publicly on the production costs for renewable diesel and jet fuel. Some techno-economic analyses (TEAs) have been conducted for certain technology pathways, and other data from specific companies can be extracted from S-1 filings when a company goes public.

In an effort to make the production costs reported here as comparable as possible, only recent sources (2008 to 2011) have been cited, and care was taken to exclude estimates with unusual assumptions such as optimistic feedstock price. However, the cost estimates are still not perfectly comparable. Variables such as plant size and co-product credits can have a significant impact on the overall production cost. Also, each pathway is at a different level of technology maturity, which contributes to the uncertainty of the cost estimates. None of the analyses cited address costs associated with carbon capture and sequestration (CCS), and none benefit from GHG emission credits.

Biodiesel

An analysis of soybean biodiesel (Tao and Aden 2009) reports a production cost of $2.55 per gallon of biodiesel ($2.48 gallon-of-gasoline-equivalent or gge) in 2007 dollars at a feedstock cost of $0.30/lb of soybean oil. The feedstock cost is by far the most significant cost of biodiesel production, accounting for about 70% of the overall production cost, and 75%-95% of the overall operational cost. Other publications confirm the significance of feedstock cost, estimating it accounts for as much as 80% of the overall production cost (Yusuf et al., 2011). A literature review in Tao and Aden (2009) revealed overall production costs ranging from $2.00-$2.50 per gallon ($1.94- $2.43/gge), although the cost years of the data points in the literature review are not provided. Current biodiesel production facilities use a caustic soda-based catalyst, sodium methoxide, which complicates product clean up. Enzymatic transesterification using a lipase biocatalyst is an attractive alternative due to reduced wastewater treatments needs, easy glycerol recovery, and absence of side reactions—although the production costs are significantly higher (Jegannathan et al., 2011). An economic analysis employing an immobilized lipase catalyst resulted in a biodiesel production cost of $8.04 per gallon ($7.81/gge), although no cost year is provided.

Hydroprocessing of Lipid Feedstock

Pearlson et al. (2013) is the only publicly-available, peer-reviewed journal article that could be found containing production cost estimates for HEFA fuels. It models diesel and jet fuel production via hydroprocessing of soybean oil for three production capacity scenarios: 31, 61, and 100 Mgy. Maximizing the process for diesel production (jet fuel and LPG are co-products) results in cost per gallon ranging from $3.82 to $4.39/gal ($3.61 to $4.15/gge). Catalytically cracking a portion of the diesel range products was modeled to maximize jet fuel production (diesel and LPG are co-products), which resulted in a jet fuel cost per gallon of $4.09 to $4.69/gal ($3.81 to $4.37/gge). No cost year is provided, although input costs for the model, such as electric power, natural gas, and purchased hydrogen, are cited in 2010 dollars.

An estimate of renewable diesel production cost via the HEFA technology pathway can be gleaned from Syntroleum Corporation financial disclosure statements. Dynamic Fuels produces renewable diesel, along with naphtha and LPG co-products, from animal fats and yellow grease. According to Syntroleum Corporation financial statements (Syntroleum 2013), the feedstock was $3.62/gal of renewable diesel produced, and operating expenses were $1.03/gal of renewable diesel produced for the quarter ending September 30, 2012 (Syntroleum 2013). Cost per gallon due to capital costs is not provided, but the Dynamic Fuels website reports the capital cost of the 75 Mgy production facility at $150 million (Dynamic Fuels 2013). Assuming a 30-year project life and 10% discount rate, we can amortize a $150 million overnight capital cost over the total amount of fuel produced over the life of the project. The resulting capital cost is $0.22/gal. Adding the capital cost per gallon to the feedstock cost and operating cost results in a production cost of about $5/gge. It should be emphasized that this cost was not reported by Syntroleum Corporation, Tyson Foods, or Dynamic Fuels. It was estimated in this study based on feedstock and operating costs provided by Syntroleum, and total capital cost (provided by Dynamic Fuels) amortized using financing assumptions common in scoping studies such as the National Renewable Energy Laboratory (NREL) techno-economic reports cited elsewhere in this section.

Algae

Davis et.al. (2011) recently published modeled costs for the co-production of diesel and naphtha from photosynthetic algae. In this analysis, both open pond and photobioreactor technologies are examined. In general, the costs for finished (hydro-treated) blendstocks range from \$9.84/gallon (\$9.30/gge) for open-pond systems to \$20.53/gallon (\$19.39/gge) in 2007 dollars for photobioreactor systems.

Algal biofuel production is a nascent technology at commercial scale. Therefore, previous analyses apply inconsistent assumptions, making comparison of results difficult. Consequently, the Department of Energy's (DOE) Bioenergy Technology Office (BETO) hosted a workshop at the University of Arizona, Tucson, from November 30–December 1, 2011, for the purpose of harmonizing assumptions used in previous algal biofuel TEAs, life cycle assessments (LCAs), and resource assessments (RAs). Due to the favorable cost estimates of open-pond systems over photobioreactor systems, only open-pond systems were considered for the harmonization exercise. Applying the harmonized assumptions resulted in a cost of \$19.60/gal (\$18.52/gge) in 2007 dollars. The increase in cost above the Davis et al. (2011) estimate is due largely to the addition of pond liners, a reduction of the baseline algae production activity from 25 g/m^2/day to 13.2 g/m^2/day, and accounting for location and seasonal variabilities (Davis et al., 2012).

Fischer Tropsch Diesel

Gasification and pyrolysis modeled costs have been most recently described in a series of papers from ConocoPhillips/Iowa State. Anex et al. (2010) estimate the production cost of diesel and gasoline produced via gasification of corn stover followed by FT synthesis at \$4.50 to \$5.00/gge in 2007 dollars, depending on the operating temperature of the gasifier. The National Energy Technology Laboratory (NETL 2009) estimates the production cost of diesel produced via gasification followed by FT synthesis as \$6.45/gallon (\$6.09/gge) in 2008 dollars. The NETL study assumes a 20% rate of return, whereas Anex et al. (2010) assume only a 10% rate of return.

Wright et al. (2008) explore distributed pyrolysis processing (on-farm and small co-op pyrolyzers) followed by centralized gasification and FT synthesis of the pyrolysis oil to FT liquids. The distributed processing scenario yields FT liquids for \$1.43 to \$1.56 per gge, although the analysis does not describe the composition of the FT liquid product. Additional separation and upgrading steps may be necessary. Additionally, transporting raw pyrolysis oil can be problematic due to its tendency toward phase separation. No cost year is provided for the estimate.

Non-Catalytic Fast Pyrolysis

Brown et al. (2011) perform a techno-economic analysis of a non-catalytic fast pyrolysis process that produces gasoline from corn stover for a selling price of $2.96/gal in 2007 dollars. Biochar and pyrolysis gas are also produced, but are consumed in the overall process for heat generation. The final price is most sensitive to bio-oil yield, followed by feedstock price. The same research group analyzed two fast pyrolysis pathways, one with on-site hydrogen generation for fuel upgrading, the other relying on merchant hydrogen (Wright et al., 2010). Results show selling prices of $3.09/gge for the on-site hydrogen production scenario and $2.11/gge with purchased hydrogen. Selling prices in 2007 dollars for pioneer plants are $6.55/gge for the on-site hydrogen production and $3.41/gge with purchased hydrogen.

Catalytic Fast Pyrolysis

The National Academy of Sciences (NAS) reviewed cost estimates made by Wright et al. (2009) and catalytic fast pyrolysis company KiOR, based on KiOR's Form S-1 filing to the Securities and Exchange Commission (NAS 2011). Using its own feedstock cost and financing assumptions, the NAS updates the estimated selling prices of $2.10/gge for Wright et al. (2009) and $3.24/gge for the KiOR catalytic fast pyrolysis process, although most details of the analysis are not provided, and it is unclear what year the costs are indexed.

Hydropyrolysis

Marker et al. (2012) modeled an integrated hydropyrolysis and hydroconversion process for gasoline and diesel production, called IH^2, developed by the Gas Technology Institute (GTI). Biomass is pyrolyzed in the presence of hydrogen to gas and liquid products. The gas phase proceeds to a hydro-conversion stage, which removes oxygen, resulting in deoxygenated gasoline and diesel products. The production cost of the finished fuel is \$1.60/gal in 2007 dollars. A gge production cost is not provided, although the product mix is reported as approximately 75% gasoline and 24% diesel, which results in a gge production cost of approximately \$1.56 in 2007 dollars.

Biojet Fuel

A review of biojet fuel cost produced several estimates, but only one peer-reviewed journal article. Agusdinata et al. (2011) estimates the cost in 2007 dollars of producing biojet fuel via gasification and FT synthesis at \$4.00/gge from corn stover, \$5.50/gge for switchgrass, and about \$5.80/gge for short rotation woody crops (SRWCs). Several estimates found in trade journals and industry blogs are consistent with this cost range of \$4 - \$6, although these sources are not reviewed and do not provide supporting material. Agusdinata et al. (2011) also estimates the cost of producing biojet fuel from algae in open ponds to be about \$17/gge.

Renewable Identification Numbers and the Renewable Fuel Standard

The RFS program was created to ensure that transportation fuel sold in the United States contains a minimum volume of renewable fuel. The EPA is responsible for implementing the RFS regulations. The obligation to meet the minimum volume falls on fuel blenders.

Renewable Identification Numbers (RINs) are the mechanism used by the EPA to track volumes of renewable fuel and verify blenders are meeting minimum blending requirements. Each volume of renewable fuel produced has an RIN attached to it. Each fuel blender must acquire enough RINs to cover its share of the mandate. RINs are tradable and can be purchased and sold. Therefore, a blender can meet RFS requirements by actually blending the mandated amount of biofuels, or by purchasing RINs from other fuel blenders who blend more biofuels than required and have excess RINs (FAPRI-MU 2009).

The RFS and RIN trading system do not affect the production cost of renewable fuels, but are worth mentioning here because they can serve as economic drivers to make renewable fuels competitive with petroleum-derived fuels (NABC 2012a) and aid in initial market penetration.

DEMAND

Diesel

Distillate fuel oils, a category that includes diesel and heating oil, are a general classification for one of the fractions obtained from petroleum distillation. These oils are used in all sectors of the U.S. economy and rank second behind gasoline as the most consumed liquid fuels (EIA 2012c). Diesel fuel (often referred to as No. 2 diesel fuel) is defined by EIA as a fuel that has distillation temperatures of approximately 500°F at the 10% recovery point and no more than 640°F at the 90% recovery point; it meets the specifications defined in ASTM Specification D 975.[2] No. 1 diesel fuel is a light distillate fuel oil that has distillation temperatures of no more than 550°F at the 90% point and meets the specifications defined in ASTM Specification D 975. Both fuels are used in high-speed diesel engines found in trucks, buses, automobiles, and locomotives, as well as farm and construction equipment. No. 1 diesel exhibits a much lower cloud point than No. 2 diesel, so it is used neat, or it is blended with No. 2 in winter months. Heating oil (often referred to as No. 2 fuel oil) is used for domestic heating and for moderate capacity commercial/industrial buildings.

EPA defines it as a distillate fuel oil that has a distillation temperature of up to 640°F at the 90% recovery point and it meets the specifications defined in ASTM Specification D 396.[2] No. 1 fuel oil is a light distillate fuel oil that has distillation temperatures of up to 400°F at the 10% recovery point and no more than 550°F at the 90% point; it meets the specifications defined in ASTM Specification D 396. It is used primarily as fuel for portable outdoor stoves and portable outdoor heaters and is commonly referred to as kerosene. More than three-quarters of distillate fuel sales are used primarily for transportation: on-highway (by trucks, buses, and automobiles); railroad; vessel bunkering; and construction, farm, and military equipment (Figure 1). About 6% of distillate sales are for residential heating purposes, concentrated during the winter months.

Table 8 in the Appendix provides a detailed breakdown of the different types of fuels used in each sector, as well as their volumes during the period 2007-2012.

On-highway motor vehicles consume about 64% of distillate fuel oils, namely diesel, with freight trucks using most of the fuel. The states of Texas and California are the largest consumers of diesel in the country, accounting for about 20% of all sales (Figure 2).

Diesel consumption by on-highway motor vehicles follows population distribution: the top consuming states are also the most populated. Table 9 in the Appendix provides information on the historic use of No. 2 diesel fuel by state.

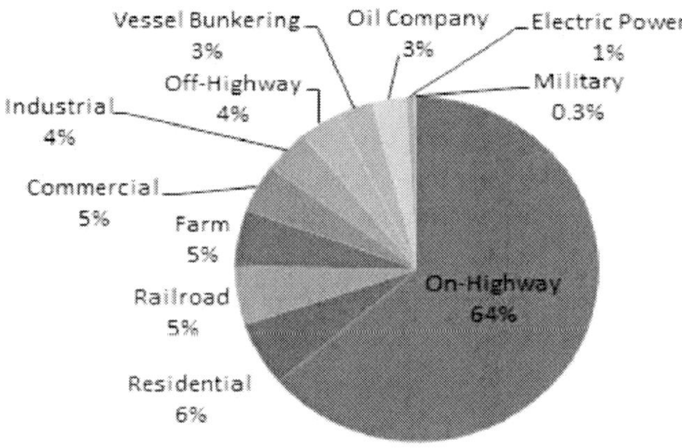

Data source: EIA 2013b.

Figure 1. Sales of distillate fuel oil by end use, 2012.

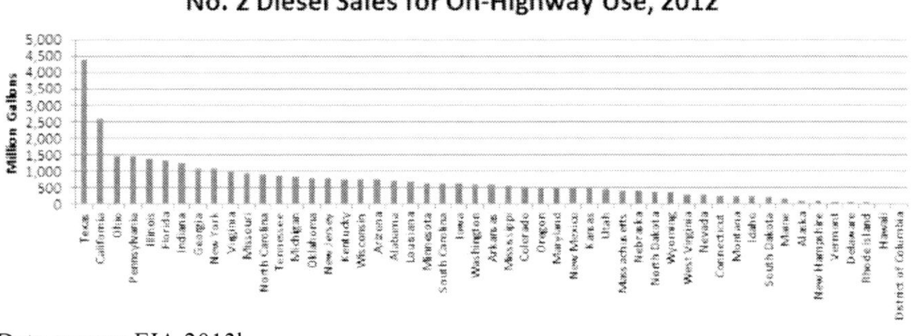

Data source: EIA 2013b.

Figure 2. No. 2 diesel sales for on-highway use by state, 2012.

Total distillate consumption was on the rise during the 1990s, and it reached a plateau during 2003-2007 (Figure 3). The growth was mainly due to increased use of diesel for transportation. From 2007-2009, low demand and a weak economy contributed to a downward trend. In 2010, this reversed to an upward trend, which experts believe is due to growth in manufacturing, usually associated with trucking demand.

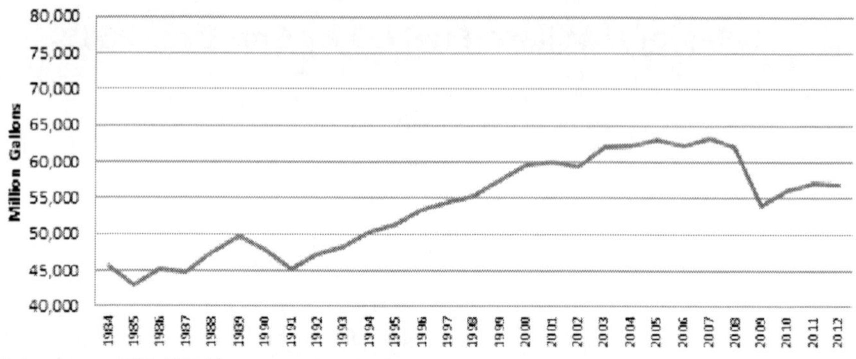

Data source: EIA 2013b.

Figure 3. Total U.S. distillate retail deliveries, 1984 – 2012.

Diesel consumption is projected to continue to grow, as illustrated in Figure 4. EIA notes that this growth results from both an increase in industrial output that leads to more fuel use by heavy trucks and an expansion of light-duty diesel vehicle sales to meet more stringent Corporate Average Fuel Economy (CAFE) standards.[3]

As shown in Figure 5, freight trucks are expected to continue to dominate diesel consumption. Freight rail fuel use is expected to increase slightly while the consumption of diesel by the remaining market segments (such as commercial light trucks, shipping, school buses, etc.) remains almost flat over the years (Figure 6).

An exception is the transit bus segment, which is projected to decrease its diesel use because of an increase in natural gas use. Historical diesel share of the transit bus segment declined relative to natural gas transit buses between 1995 and 2008, and this tradeoff is expected to continue.

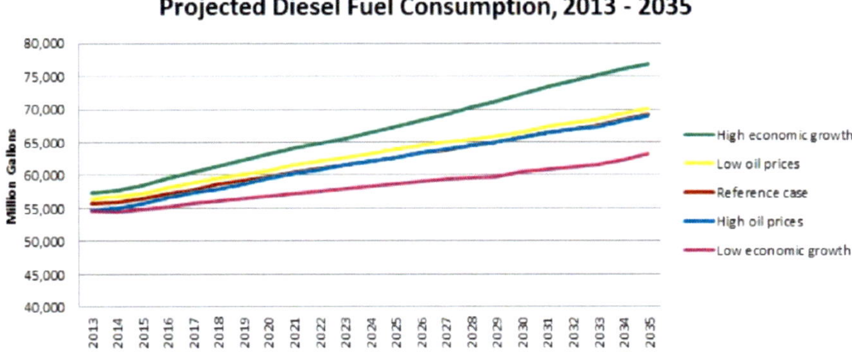

Data source: EIA 2011.

Figure 4. Projected diesel fuel consumption, 2013 – 2035.

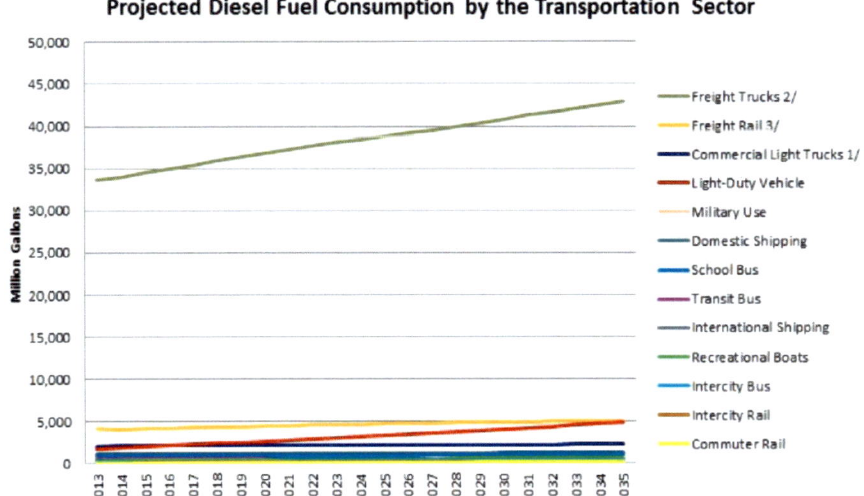

Data source: EIA 2011, Reference Case.

1/ Commercial trucks from 8,500 to 10,000 pounds.

2/ Does not include military distillate. Does not include commercial buses. 3/ Does not include passenger rail.

Figure 5. Projected diesel fuel consumption by the transportation sector.

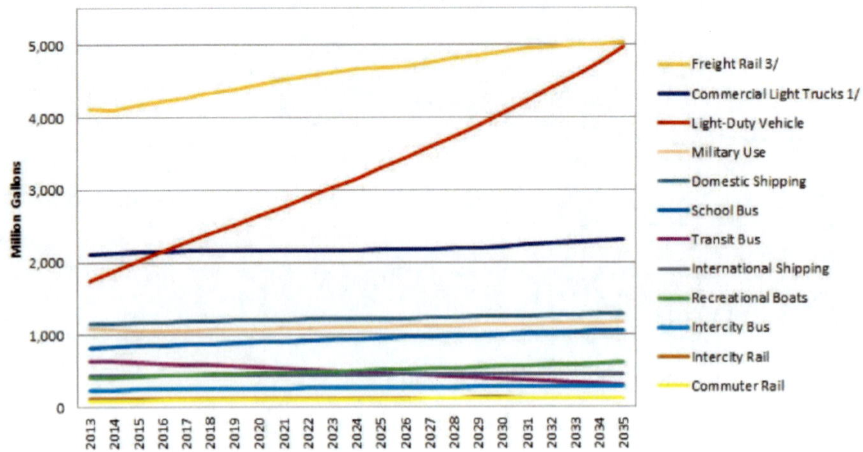

Data source: EIA 2011, Reference Case.
1/ Commercial trucks from 8,500 to 10,000 pounds. 3/ Does not include passenger rail.

Figure 6. Projected diesel fuel consumption by the transportation sector excluding freight trucks.

Biodiesel

Biodiesel consumption was on the rise from 2005-2009, plummeted during 2010 due to low demand resulting from the economic downturn and the expired $1.00 per gallon blender tax credit, and was back up in 2011 mainly in response to RFS2 and the reinstated tax credit (Figure 7). Biodiesel is distributed through fueling stations nationwide. Low-level biodiesel blends such as B2 and B5 are used safely in any compression-ignition engine designed to use diesel fuel. There are several hundred major fleets in the United States that use biodiesel. These include municipal bus fleets; national park trucks and buses; federal, state, and local government fleets; school buses; and many commercial businesses, such as public utilities and refuse haulers (Hart Energy Consulting 2010). Off-road applications (tractors, boats, and electrical generators) also use biodiesel. EIA projects an annual growth of 7% for biodiesel consumption by 2035 (EIA 2011).

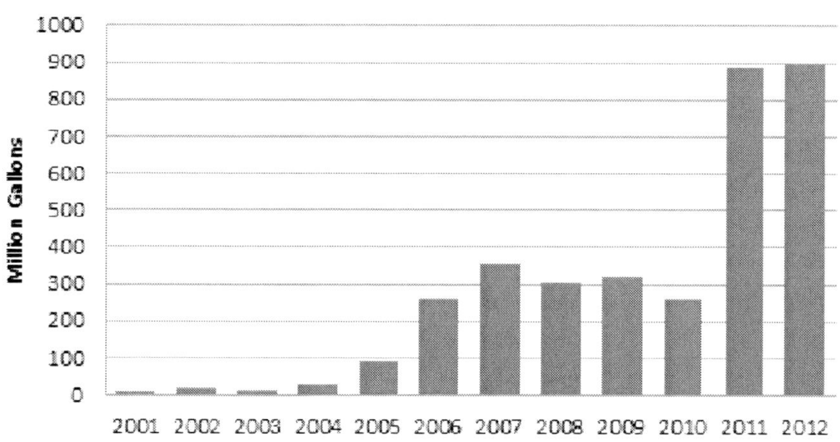

Data source: EIA 2013a.

Figure 7. Historic biodiesel consumption, 2001 – 2012.

Jet Fuel

Jet fuel is a type of aviation fuel designed for use in commercial and military aircrafts powered by gas-turbine engines. It is the third-most used fuel in the country after gasoline and diesel. EIA defines jet fuel as a kerosene-based product having a maximum distillation temperature of 400°F at the 10% recovery point and a final maximum boiling point of 572°F and meeting ASTM Specification D 1655 (JET A and JET A-1) and Military Specifications MIL-T-5624P and MILT-83133D (Grades JP-5 and JP-8).[4]

Jet fuel consumption was on the rise during the 1980s and 1990s and it reached a plateau during 2000 - 2007 (Figure 8). Since 2007, similar to diesel consumption, jet fuel consumption began a downward trend due to low demand and a weak economy. States consuming the most jet fuel include California, Texas, New Jersey, Florida, and Illinois (Figure 9); they are home to some of the busiest airports in the United States as well as large military bases.

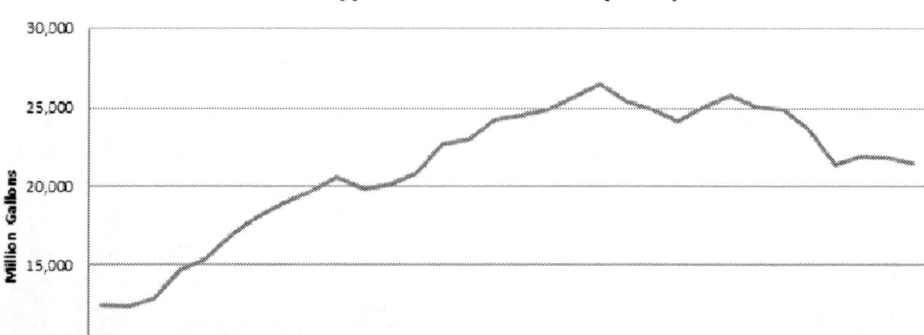

Data source: EIA 2013c.

Figure 8. Historic kerosene-type jet fuel consumption, 1981 – 2012.

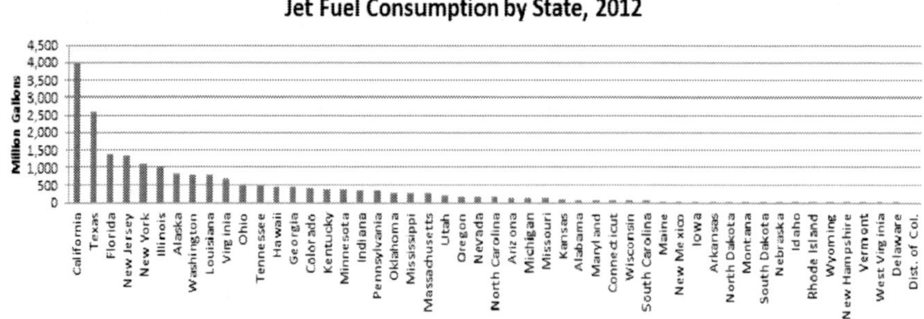

Data source: EIA 2013d.

Figure 9. Jet fuel consumption by state, 2012.

ETA projects that jet fuel consumption by commercial carriers will continue to grow over the next years, the rate of which will depend on economic growth and oil prices, whereas jet fuel consumption by the military will remain flat (Figure 10 and Figure 11).

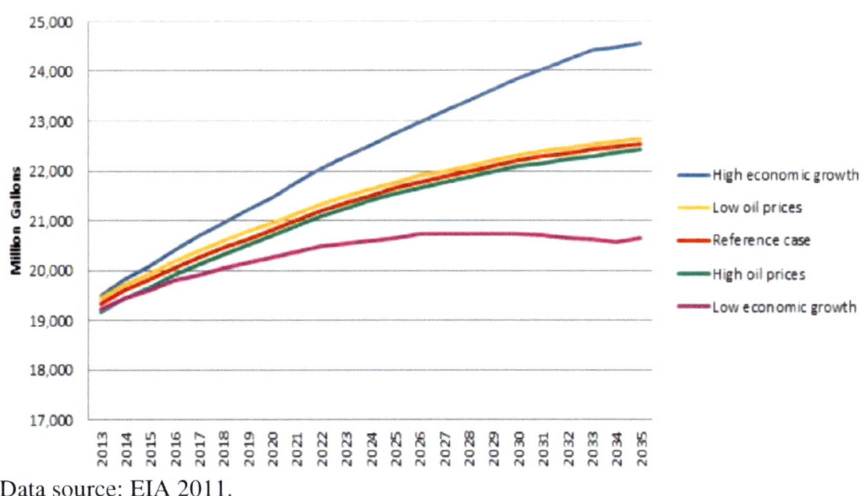

Data source: EIA 2011.

Figure 10. Projected commercial jet fuel consumption, 2013 – 2035.

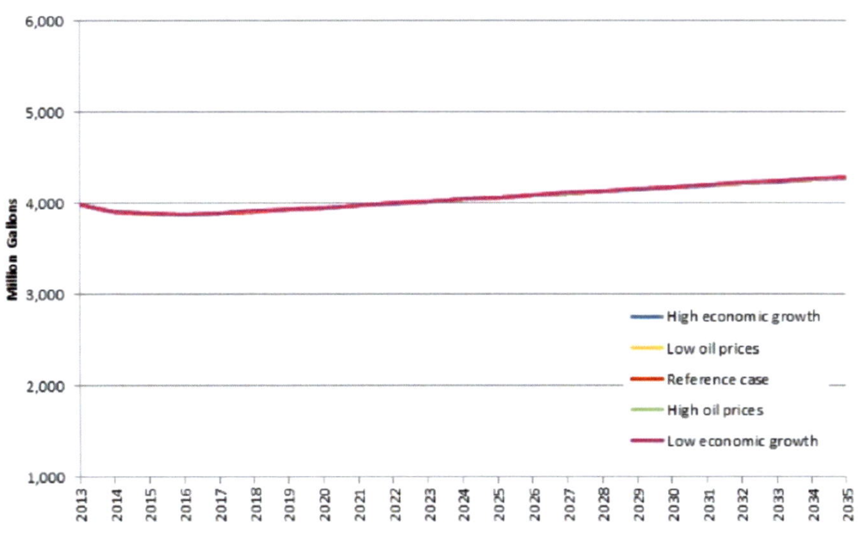

Data source: EIA 2011.

Figure 11. Projected military jet fuel consumption, 2013 – 2035.

FEEDSTOCK ASSESSMENT

Various biomass resources can be used for the production of diesel and jet fuel substitutes, as established earlier. These include lignocellulosic material such as wood waste, crop residues, and dedicated energy crops, as well as lipid feedstock such as vegetable and waste oils, animal fat, and algae. A recent study by the U.S. Department of Energy's Oak Ridge National Laboratory (U.S. DOE 2011), the Billion Ton Study (BTS), estimated the amount of currently available and potential biomass resources in the conterminous United States. The study included the following biomass categories:

1. Forest Biomass and Wood Waste

- Forest residues (logging residues and thinnings) from integrated forest operations from timberland
- Other removal residue[5]
- Thinnings from other forestland[6]
- Unused primary and secondary mill processing residues[7]
- Urban wood wastes (woody component of municipal solid waste [MSW] and construction and demolition [C&D] wood)
- Conventionally sourced wood[8]

2. Agricultural Biomass and Waste Resources

- Crop residues from the major grain-producing crops (corn, wheat, barley, oats, and sorghum)
- Secondary agricultural processing residues (sugarcane trash and bagasse, cotton gin trash and residues, soybean hulls, rice hulls and field residues, wheat dust and chaff, orchard and vineyard prunings)
- Waste or tertiary resources (e.g., manures, waste fats, and greases)

3. Energy Plantations (perennial grasses, trees, and annual crops).

The BTS makes use of POLYSYS, an agricultural policy model, and data from the United States Department of Agriculture (USDA) to estimate supply/cost curves for each feedstock by county for 2012-2030. Two scenarios are used: baseline and high yield. The baseline scenario assumes a continuation of USDA's 10-year forecast for the major food and forage crops, as well as a continuation in trends toward no-till and reduced cultivation.

Energy crop yields assume an annual increase of 1% due to experience in planting and additional R&D. Forest residues are estimated using resource cost analysis with data from the USDA's Forest Service. The high-yield scenario assumes a greater proportion of corn in reduced and no-till cultivation and increased corn yields to about double the current rate of annual increase. The energy crop productivity increases at 2%, 3%, and 4% annually, not only due to experience in planting, but also to more aggressive implementation of breeding and selection programs. No high yield scenario was evaluated for forest resources except for the woody crops.

The forest resources are available over a wider price range than the agricultural resources–$20 per dry ton or less to $100 per dry ton or less for forest resources vs. $40 per dry ton or less to $60 per dry ton or less for agricultural resources–with increasing quantities at higher prices. Over the estimated price range, forest resource quantities vary from about 33 to 142 million dry tons currently (Table 2). Due to data limitations, there is little estimated change over the next 20 years. Another study looked at the currently available forest resources in the country, regardless of price, and estimated that there are about 92 million dry tons of woody biomass available in the country today (Milbrandt 2005).[9] A study by NAS reports a similar amount of currently available woody biomass, about 110 million dry tons, and projects that about 124 million dry tons could be available in 2020 (NAS 2009).

Table 2. Summary of Potential Forest Biomass and Wood Wastes in the BTS, 2012

Feedstock ($ per dry ton)	<$20	<$30	<$40	<$60	<$80	<$100
			Million dry tons			
Other Removal Residue	4.4	12	12	12	12	12
Composite Operations	9.5	30	36	40	42	43
Without Federal Land	8.3	26	31	35	36	37
Treatment Thinnings, Other Forestland	0	0	0	3.2	6.4	6.4
Without Federal Land	0	0	0	1.8	3.6	3.6
Mill residue, unused primary	1.3	1.3	1.3	1.3	1.3	1.3
Mill residue, unused secondary	6.1	6.1	6.1	6.1	6.1	6.1
Urban Wood Waste – C & D	4.4	11	14	22	22	22
Urban Wood Waste – MSW	7.7	8.7	9.2	10	10	10
Conventional Pulpwood to Energy*	0	0	0	1.5	19	40
Total – All Land	**33**	**70**	**79**	**97**	**119**	**142**
Total – Without Federal Land	*32*	*66*	*75*	*90*	*111*	*133*

Source: U.S. DOE 2011.

Under the baseline scenario, agricultural resource quantities vary from about 59-162 million dry tons currently to about 126-265 million dry tons in 2030. About two-thirds of this quantity comes from crop residue and the rest is from various agricultural processing residues and wastes. Under the high yield scenario, the quantities vary from about 115-244 million dry tons currently to about 284-404 million dry tons in 2030 (Table 3). As a reference, Milbrandt (2005) reports about 157 million dry tons of crop residues available in 2002. This amount is one-third of the total crop residues, accounting for a portion that needs to be left on the field for soil protection, grazing, and other agricultural activities. NAS (2009) reports a lower amount of currently available agricultural resources[10]--about 106 million dry tons—and projects that about 148 million dry tons of agricultural resources could be available in 2020.

Table 3. Summary of Baseline and High Yield Scenarios in the BTS — Agricultural Residues and Waste Resources

	<$40 per dry ton				<$50 per dry ton				<$60 per dry ton			
Feedstock	2012	2017	2022	2030	2012	2017	2022	2030	2012	2017	2022	2030
	Million dry tons											
Baseline												
Corn	19	32	42	65	73	93	108	129	85	106	120	140
Wheat	6.7	7.8	9.1	12	18	22	26	31	23	26	31	36
Barley, Oats, Sorghum	1.0	1.3	1.6	2.9	2.4	2.5	2.4	3.6	2.8	2.7	2.6	3.7
Total primary residue	**27**	**41**	**52**	**80**	**94**	**117**	**136**	**164**	**111**	**135**	**154**	**180**
Secondary residues & wastes												
Rice field residue	6.5	6.9	7.4	8	6.5	6.9	7.4	8	6.5	6.9	7.4	8
Rice hulls	1.5	1.6	1.7	1.7	1.5	1.6	1.7	1.7	1.5	1.6	1.7	1.7
Cotton field residue	4.2	5.3	5.9	6.7	4.2	5.3	5.9	6.7	4.2	5.3	5.9	6.7
Cotton gin trash	1.4	1.6	1.7	1.8	1.4	1.6	1.7	1.8	1.4	1.6	1.7	1.8
Sugarcane residue	1.1	1.1	1.1	1.1	1.1	1.1	1.1	1.1	1.1	1.1	1.1	1.1
Orchard and vineyard prunings	5.7	5.6	5.5	5.5	5.7	5.6	5.5	5.5	5.7	5.6	5.5	5.5
Wheat dust	0.6	0.6	0.6	0.6	0.6	0.6	0.6	0.6	0.6	0.6	0.6	0.6
Animal manures	12	13	16	20	29	34	41	56	30	35	43	59
Animal fats	0	0	0	0	0	0	0	0	0	0	0	0
Total secondary residues & wastes	**33**	**36**	**40**	**46**	**50**	**56**	**65**	**82**	**51**	**58**	**67**	**84**
Total baseline	**59**	**77**	**92**	**126**	**143**	**174**	**201**	**245**	**162**	**192**	**221**	**265**
High-yield scenario												
Corn stover	71	132	157	221	143	200	228	264	153	209	234	271
Wheat Straw	9.8	12	13	16	60	35	38	42	35	39	42	46
Barley, Oats, Sorghum	1.5	1.5	1.4	1.7	3.6	3.4	2.8	3.1	4.0	3.6	2.9	3.0
Total primary residue	**83**	**146**	**171**	**238**	**176**	**239**	**269**	**309**	**193**	**252**	**279**	**320**
Total high-yield	**115**	**182**	**210**	**284**	**226**	**295**	**334**	**391**	**244**	**310**	**346**	**404**

Source: U.S. DOE 2011.

Energy crops considered in the BTS include switchgrass, miscanthus, sugarcane, sorghum, poplar, willow, eucalyptus, and southern pine. The study estimates that between 3.7 and 101 million dry tons of energy crops could be available under the baseline scenario in 2017 (Table 4). This wide range is due to uncertainty in crop yield and the rate of industry's development. The high yield scenario is simulated at three levels–2%, 3%, and 4% increase in annual crop productivity. Over the estimated price range, the quantity of energy crops varies between 13 and 180 million dry tons in 2017 to about 69 and 799 million dry tons in 2030. NAS (2009) reports that with current technologies and agricultural practices, about 104 million tons of energy crops could be produced today. Advanced technologies and practices could lead to increased production, potentially up to 164 million tons in 2020.

Table 4. Summary of Baseline and High Yield Scenario Availability of Energy Crops in the BTS

	<$40 per dry ton			<$50 per dry ton			<$60 per dry ton		
Feedstock	2017	2022	2030	2017	2022	2030	2017	2022	2030
Baseline scenario					*(Million dry tons)*				
Perennial grasses	3.0	12	30	41	77	129	90	188	256
Woody crops	0.0	0.0	0.1	0.9	40	67	5.7	84	126
Annual energy crops	0.7	1.8	4.2	3.8	7.3	14	5.0	10	19
Total	**3.7**	**14**	**34**	**46**	**124**	**210**	**101**	**282**	**400**
High-Yield (2% annual growth)									
Perennial grasses	11	43	57	67	152	239	122	253	319
Woody crops	0.0	0.1	4.2	1.9	78	127	10	145	207
Annual energy crops	1.6	4.1	7.4	5.5	8.7	12	6.9	11	15
Total	**13**	**47**	**69**	**75**	**239**	**378**	**139**	**409**	**540**
High-Yield (3% annual growth)									
Perennial grasses	24	71	107	85	213	329	138	296	390
Woody crops	0.0	1.5	43	9.3	101	186	14	168	251
Annual energy crops	2.4	6.6	11	6.2	10	14	8.0	12	18
Total	**26**	**79**	**162**	**101**	**324**	**520**	**160**	**476**	**658**
High-Yield (4% annual growth)									
Perennial grasses	35	100	202	106	270	406	154	338	462
Woody crops	0.1	5.3	45	12	118	199	16	212	315
Annual energy crops	3.4	9.0	14	6.8	11	18	9.4	14	22
Total	**39**	**114**	**261**	**124**	**399**	**622**	**180**	**564**	**799**

Source: U.S. DOE 2011.

FOG is another resource category for the production of diesel and jet fuel substitutes. The United States produced about 18 million tons of FOG in 2010 (Table 5), which could theoretically be converted to over 5 billion gallons of biodiesel or renewable diesel. Another 3.7 million tons of oils were imported (mainly edible oils such as canola, palm, and coconut). Edible oil imports have increased from 13% of domestic edible oil production in 1998 to 27% in 2010 as a result of the shift from soy to low polyunsaturated oils (because of negative health effects of trans-fatty acids and rising U.S. food demand). However, not all of these resources are available for energy use. Roughly 85% were used for edible (baking or frying fats, margarine, salad or cooking oil) and inedible products (fatty acids, animal feed, methyl esters, etc.). In addition, about 1.5 million tons of edible oils (mostly soy) and 900,000 tons of animal fats were exported in 2010 (USDA 2012; U.S. Census Bureau M311K). The 1.1 billion gallon biodiesel production in 2011 consumed over 4 million tons of FOG, which is a significant increase from the 1.1 million tons consumed in 2010.

Table 5. U.S. Production of Fats, Oils, and Greases in 2010

Sources	Production ('000 tons)
*Oils**	
Corn	1,258
Cottonseed	408
Peanut	82
Canola	576
Safflower	32
Soybean	9,518
Sunflower	314
Tall oil, crude**	672
Vegetable foot**	178
Fats	
Lard*	425
Edible tallow*	913
Inedible tallow**	1,650
Poultry fat**	709
Yellow grease**	702
Other grease**	660
Total Current Supply	18,092
Oils and Fats Import*	3,687

* Data Source: USDA, Oil Crops Yearbook 2011.
** Data Source: U.S. Census Bureau, M311K.

Algae are a potential aquatic oil crop, but may also yield carbohydrates that can be converted to sugar. This feedstock has received increased attention in recent years. Given the right resources– suitable climate, availability of water, CO2 and other nutrients – algal oil productivity can be quite high. A recent study suggests that under current technology, microalgae have the potential to generate 58 billion gallons of oil per year, equivalent to 48% of current U.S. petroleum imports for transportation (Wigmosta et al., 2011). The authors emphasize that this level of production requires 5.5% of the land area in the conterminous U.S. and nearly three times the water currently used for irrigated agriculture. The intensive water use for growing algae can be addressed in a sustainable manner by using low-quality water with few competing uses, such as brackish/saline groundwater, "co-produced water" from oil and natural gas wells, and wastewater discharged from domestic, industrial, and agricultural activities. Therefore, algal technology need not put additional demand on freshwater supplies.

Figure 12 illustrates the algal oil productivity at different locations in the United States. The pattern shows a strong linkage to climate and topography– locations with warm temperatures and flat terrain are most productive. The southern portions of the country exhibit the highest production rates ranging from 6,000 to 8,221 liters/ha/year (between 540-740 gallons/acre/year) of potential biofuel production. These areas are characterized with relatively warm temperatures year-round and longer hours of solar insolation in comparison to northern locations. Locations in the north and at higher elevations exhibit the lowest production rates, ranging from 2,291 to 4,000 liters/ha/year (206-360 gallons/acre/year). A long winter season in these areas contributes to a shorter growing season.

Table 6 summarizes the biomass resource potential outlined above. Depending on the conversion pathway, different yields of biofuels can be achieved. Using current conversion technologies for lignocellulosic biomass, about 43.3 gallons of renewable diesel/jet fuel could be produced per dry ton via gasification/FT technology (Davis 2009) and about 65 gallons per dry ton via fast pyrolysis (PNNL 2009). For example, forest resources alone could produce about 1.4 - 6 billion gallons of renewable diesel via gasification/FT technology which could displace between 4% and 17% of current diesel consumption on highways (about 36 billion gallons of diesel fuel was used in 2011; see Figure 1 and Figure 3). Using fast pyrolysis, forest resources could yield even more "drop-in" fuels and displace between 6% and 26% of current diesel consumption. Renewable diesel and jet fuel yields from lignocellulosic biomass via biochemical conversion pathways are not readily available at this time.

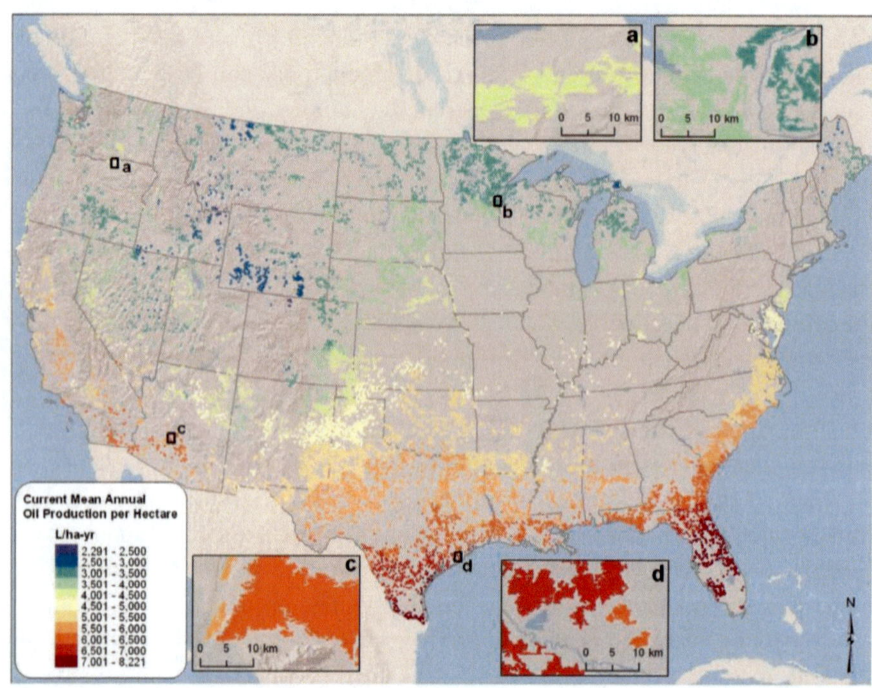

Source: Wigmosta et al., 2011.

Figure 12. Mean annual algal oil production using current technology.

On average, about 300 gallons of biodiesel, using a transesterification process, could be produced per ton of FOG. Thus, the 3.8 million tons of currently available FOG could yield about 1.1 billion gallons of biodiesel, which could displace about 3% of current diesel consumption on highways in the United States. About 245 gallons of "drop-in" fuel could be produced per ton of triacylglycerol (TAG) oil via hydroprocessing (Davis et al., 2012). Thus, about 931 million gallons of renewable diesel could be produced from the existing FOG in the United States, which could displace about 2.5% of current diesel consumption on highways in the country.

It is unrealistic to assume that all of the biomass resources could be used for the production of diesel and jet fuel substitutes. There are other competing uses for these resources. For lignocellulosic biomass, these include the production of ethanol, renewable gasoline, chemicals and allied products, as well as power generation. For FOG, competing industries include the production of edible (baking/frying fats, cooking oil, etc.) and inedible (lubricants, paints, plastics, etc.) products. However, even if a small portion of the current biomass potential is dedicated to diesel and jet fuel substitutes, they can still play a role in diversifying the country's energy portfolio. This role would be more significant if the projected future potential is realized and feedstock, such as dedicated energy crops and algae, become commercially available.

Table 6. Total Biomass Resource Potential

Feedstock	Current Potential (million tons)	Future Potential in 2022 (million tons)
Forest resources	33-142	34-150
Crop residues	59-162	92-221
Energy crops	n/a	14-282
Fats, Oils, and Greases*	3.8	4.1
Algae**	n/a	400

* Assumes 21% of FOG [about 85% of total FOG (18 million tons) is currently used for edible and inedible products, of which 6% is used in methyl esters (biodiesel)]. Future potential is estimated using population projections from the U.S. Census Bureau in 2022 (0.93 percent change) and applying the same ratio (21% of total).

** Assumes 1/5 of projected future potential.

DISCUSSION

The Congressionally mandated RFS2 goal is to use at least 36 billion gallons of biomass-based transportation fuels by 2022 (Figure 13). Of the total 36 billion gallons, 15 billion gallons are projected to come from conventional biofuel sources, such as corn ethanol, and the remaining 21 billion gallons from advanced biofuels divided into three categories: cellulosic biofuels, biomass-based diesel, and other advanced biofuels (EPA 2009, Table V.A. 2-1). Sixteen billion gallons are required to come from cellulosic biofuels (fuels, not necessarily ethanol, made from lignocellulosic biomass that also reduce GHG emissions by at least 60% relative to the gasoline or diesel fuel they displace). The contribution of biomass-based diesel to this goal can be no less than 1 billion gallons[11]: 0.81 billion gallons are projected to come from biodiesel and 0.19 billion gallons from renewable diesel. An additional 4 billion gallons of other advanced biofuels (any renewable fuel other than ethanol derived from corn that achieves 50% GHG emissions) is also mandated by RFS2.

Biomass-based diesel and jet fuel are considered to be advanced biofuels and can meet several RFS criteria. For example, renewable diesel derived from lignocellulosic biomass falls within the biomass-based diesel and cellulosic biofuels categories while jet fuel substitutes from biomass can meet the cellulosic biofuels and other advanced biofuels standards.

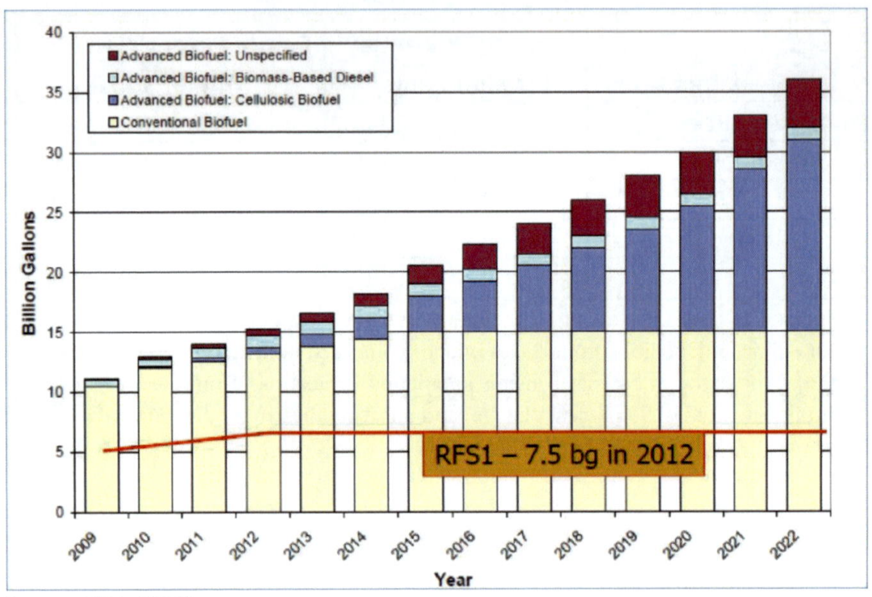

Source: EPA 2010b.

Figure 13. Renewable fuel volume requirements for RFS2 under EISA.

During 2011 and 2012, the biodiesel industry proved that it has the ability to fulfill the entire biomass-based diesel volume requirements under RFS2. The mandate calls for at least 1 billion gallons of biomass-based diesel annually and biodiesel production was reported at 1.1 billion gallons during these two years. Production in 2013 is forecasted to be even higher. In addition to the RFS mandate, another factor for the industry's growth during the past three years was the extended tax credit. This key $1.00-per-gallon blender tax credit is scheduled to expire on December 31, 2013 and it is unknown at this point whether or not it will be extended. The industry could face significant setbacks if extension does not occur. In addition, EPA proposes to keep the biomass-based diesel standard at 1.28 billion gallons next year, the same volume set for 2013, which could further jeopardize the industry's further development (EPA 2013c). The next several months will reveal the biodiesel industry's future directions as these two legislative pieces get resolved.

Biofuels Digest (2012) projects that "drop-in" fuels capacity (inducing renewable diesel and jet fuel) will reach about 1 billion gallons in 2017. This projection is based primarily on companies' statements and media releases on planned facilities. Whether or not this capacity level is reached will depend on many factors, including commercialization of proven technologies (namely gasification, FT, and pyrolysis), success of trial technologies, feedstock availability, production cost, and diesel/jet fuel demand. As mentioned earlier, currently there are three commercial plants in the United States producing renewable diesel, although one of them (KiOR) produces other renewable fuels as well such as renewable gasoline. At present, Dynamic Fuels is idle and it is unknown when the facility will resume operation again. The other two plants, KiOR and Diamond Green Diesel are in an initial phase and thus it is difficult to evaluate the performance and economics of these plants. Neste Oil is the world's largest renewable diesel producer with about 572 MGY of installed capacity worldwide (*Biofuels Digest* 2012). It operates abroad in Finland, Singapore, and the Netherlands. While Neste Oil is using hydrotreating, a common practice in petroleum refineries, other technologies such as pyrolysis (employed by KiOR) and FT (utilized by Syntroleum) have yet to demonstrate their economic viability. The next few years will likely answer many questions about the feasibility and profitability of producing renewable diesel and jet fuel through the various conversion pathways.

Based on current statistics, and as proven by the biodiesel industry, there is enough lipid feedstock to support the RFS2 mandate. Today, roughly half of the biodiesel in the United States is produced from soybean oil. The remaining portion consists of animal fat, used cooking oil, canola, and some other minor feedstocks such as cottonseed oil, tall oil fatty acids, and corn oil recovered from corn ethanol dry mills. The Food and Agricultural Policy Research Institute (FAPRI) projects that soybean oil production in the United States will continue to rise over the next decade, from about 9.5 million tons today to about 11 million tons in 2022 (FAPRI-ISU 2012). One of the drivers for this increase is the anticipated volume required to meet the biofuels quota under the RFS2. Even if half of the projected soybean potential is used for biodiesel/renewable diesel production it would be sufficient to meet the mandate. Higher output could be achieved through advanced technologies for increasing oil supply (increasing crop yield and seeds' oil content, as well as recovering sewage and trap greases) and production of new feedstock such as oil from camelina (*Camelina sativa* L.) and field pennycress (*Thlaspi arvense*). Moreover, if algal oil becomes commercially available in the next 5-10 years, the biodiesel and renewable diesel production potential would increase substantially. With regard to the lignocellulosic feedstock, enough material is projected to be available to support the RFS mandate of 21 billion gallons from advanced biofuels. Crop and forest residues in 2022 are projected to be between 126 - 371 million tons. Assuming a conversion via fast pyrolysis (with an average yield of 65 gallons/dry ton), this amount could yield about 8-24 billion gallons of biomass-based diesel/jet fuel. This potential could be larger if the conversion technologies achieve higher yield and if additional feedstock such as dedicated energy crops become available. The fuel produced from these resources would qualify under any of the advanced biofuels categories: cellulosic biofuels, biomass-based diesel, or other advanced biofuels. However, there will be competition for lignocellulosic feedstock with the ethanol industry and renewable gasoline producers to meet the RFS mandate, so it is unclear what share the renewable diesel/jet fuel would have in the total biofuels contribution. Ultimately, it will depend on the rate of commercialization of these technologies as well as the transportation market demands. The costs for producing renewable diesel and jet fuel are not well known and include a high degree of uncertainty. Hydroprocessing of FOG is being done commercially by Dynamic Fuels, Diamond Green Diesel, and internationally by Neste Oil. The KiOR technology is at an initial commercial scale. Other technology routes are not yet commercial and publicized cost estimates vary widely. As these technology pathways mature and become more widespread, more specific information regarding their economics will be available, which would enable a more detailed analysis and performance comparison. Based on Syntroleum Corporation financial disclosure statements, we estimate the production cost for renewable diesel by Dynamic Fuels to be about $5/gge as of September 2012. As a reference, the average production cost of petroleum-derived diesel in the United States is about $2.87 ($2.59/gge) as of October 2013 (EIA 2013e).[12] As another reference, production cost for biodiesel is in the range of $2.00-$2.50 per gallon ($1.94-$2.43/gge). However, it should be noted that this range was documented in 2009; thus it is likely that biodiesel production cost is higher today given the increase in soybean and other relevant feedstock prices during the past several years (USDA 2013). As a reminder, this is the cost of production (feedstock and processing), not the price paid at the pump by the consumer. The price at the pump also includes distribution expenses, taxes, etc. Biodiesel faces some technical limitations which may direct future industry decisions toward renewable diesel. While the lower-level biodiesel blends (B20 and below) can be used in traditional diesel vehicles without engine modifications, higher-level blends may require engine modifications and other usage considerations. More importantly, biodiesel is currently transported from the production sites to petroleum terminals where it is blended with petroleum fuels, via truck or rail and occasionally, by barge. These modes are much more expensive than transportation by pipeline, which is used for most petroleum fuels. The potential for biodiesel to contaminate jet fuel is preventing widespread pipeline transport. Hydrocarbon renewable diesel is likely to be easily transported in existing pipelines already utilized for petroleum-based fuels.

Renewable diesel and jet fuels produced by hydroisomerization of lipid feedstocks, gasification followed by FT synthesis, or biochemically using the approaches by Amyris, LS9, and other companies will consist entirely of paraffinic hydrocarbons (also known as alkanes). These are completely miscible with conventional fuels and, from a purely combustion standpoint, they could be blended at any level or even used in neat form as long as they meet the requirements of the respective fuel standard specifications (ASTM D975 for diesel fuel and ASTM D1655 for jet fuel). For both diesel and jet applications, it is likely that fuel additives will be required to meet specifications for lubricity, conductivity, stability, and other properties. All of these additives are also commonly used in petroleum-derived fuels. However, fuel system elastomeric components for both ground transport and aviation have been conditioned over time in a significant level of aromatic compounds. When exposed to a zero aromatic fuel such as the renewable diesel and jet fuels described above, aromatics can be extracted from the elastomers, causing them to shrink and leading to fuel system leaks. To avoid this scenario in aviation applications, ASTM International has developed standard D7566 "Standard Specification for Aviation Turbine Fuel Containing Synthesized Hydrocarbons," which allows no more than 50% volume paraffinic kerosene in jet fuel. The blended fuel is also required to contain a minimum of 8% volume aromatics. A similar situation could be envisioned for elastomeric components in diesel engine fuel systems, but to date, ASTM has not specified a minimum required aromatic level or a maximum allowable blend level of paraffinic components.

Other renewable diesel fuels, such as those produced by fast pyrolysis and related thermochemical processes, will likely contain significant levels of aromatics, and perhaps low levels of oxygenates. More data on the composition of these fuels will be required in order to assess any possible limit on blend level. For widespread application, biodiesel blending is limited to

20% volume due to lack of engine manufacturer approval and limited data on compatibility with infrastructure for higher blends.

As noted earlier, on-highway motor vehicles consume about 64% of the diesel fuel in the country, with freight trucks being the largest consumer in this category. Therefore, the trucking industry could provide a strong business opportunity for biomass-based diesel producers. The American Trucking Association (ATA) is supportive of alternative fuels, evident from the fact that it currently supports biodiesel use in blends up to 5%. The organization highlights that renewable diesel has not been subjected to rigorous on-road fleet testing; however, preliminary information indicates that renewable diesel may have advantages over biodiesel for the end-user (ATA 2012). These advantages may include a higher energy content and better cold weather performance compared to biodiesel. Regardless, for any biomass-derived diesel to be a success among the trucking companies, the fuel price has to be equal to or less than petroleum-based diesel. Table 7 illustrates the top 25 U.S. trucking companies in 2012. By far, the two major package delivery companies UPS and FedEx lead the ranking in revenue. The e-commerce boom in recent years is the main contributor to the success of these companies.

As illustrated earlier, diesel consumption by light-duty vehicles in the United States is relatively low, but it is projected to double over the next decade. The fuel is not as popular among U.S. passenger-car drivers as it is in Europe, but diesel vehicle sales show consistent double-digit increases over the last two years (*PR Newswire* 2012). Whether diesel cars remain niche vehicles in the U.S. market or enter the mainstream will depend on government policies, consumer demand, and fuel prices. The recently issued new CAFE standard of 54.5 mpg by 2025 is expected to have a positive effect on future clean diesel car sales. The Diesel Technology Forum (DTF), a non-profit educational organization dedicated to raising awareness about the economic importance and essential uses of diesel engines, issued a statement welcoming the new standards: "Because clean diesel autos are 20 to 40 percent more efficient than gasoline vehicles, diesel will be a major player in the nation's effort to achieve the new mileage standards" (*Time* 2012, 8). Demand for gasoline is higher in the United States and fuel taxes favor gasoline, which makes gas less expensive. It is exactly the opposite in Europe where the fuel tax structure favors diesel (Figure 14). The U.S. diesel passenger vehicle market may expand as consumers are drawn more to diesels because they're economical, offering greater fuel efficiency than comparable gas-powered cars. Moreover, if the U.S. fuel prices take off more abruptly than analysts predict, we could see a deeper penetration of diesel vehicles.

Table 7. Top 25 U.S. Trucking Companies in 2012

2012 Rank	Parent Company	Primary Service	Subsidiary Publics/Services & Comments	Public/Private	2011 Annual Revenue ($ million)	2012 Annual Revenue ($ million)	2011-2012 Percent Change
1	UPS	Parcel	*UPS Ground, UPS Freight*	Public	$24 752	$25 692	3.80%
2	FedEx	Parcel	*FedEx Ground, FedEx Freight, FedEx Custom Critical*	Public	$14 149	$15 416	9.00%
3	J.B. Hunt Transport Services	TL	*Truckload, Dedicated Contract Services, Integrated Capacity Solutions, Intermodal*	Public	$4 527	$5 055	11.70%
4	YRC Worldwide	LTL	*YRC Freight, YRC Regional*	Public	$4 869	$4 851	-0.40%
5	Con-way	LTL	*Con-way Freight, Con-way Truckload*	Public	$3 729	$3 898	4.50%
6	Swift Transportation	TL	*Truckload, Dedicated, Intermodal*	Public	$3 334	$3 493	4.80%
7	Schneider National	TL	*Schneider National, Schneider National Bulk Carriers, Schneider Intermodal*	Private	$3 170	$3 262	2.90%
8	Landstar System	TL	*BCO, TBC, Intermodal*	Public	$2 789	$2 952	5.90%
9	Old Dominion Freight Line	LTL	*Fastest-growing LTL carrier in Top 50 List*	Public	$1 883	$2 110	12.10%
10	Werner Enterprises	TL	*One-way Truckload, Dedicated, Cross-Border*	Public	$2 003	$2 036	1.70%
11	Arkansas Best	LTL	*ABF Freight System, Truck Brokerage, Panther Expedited*	Public	$1 756	$1 900	8.20%
12	U.S. Xpress Enterprises	TL	*Arnold Transportation subsidiary merged with LinkAmerica in December 2012*	Private	$1 670	$1 730	3.60%
13	Estes Express Lines	LTL	*Largest privately held LTL carrier*	Private	$1 640	$1 754	6.90%
14	Prime	TL	*Largest temperature-controlled carrier*	Private	$1 206	$1 372	13.80%
15	R & L Carriers	LTL	*R+L Carriers, R+L Truckload Services*	Private	$1 207	$1 230	3.00%
16	C.R. England	TL	*National, Regional, Dedicated, Intermodal, Mexico*	Private	$1 005	$1 146	14.10%
17	Greatwide Logistics	TL	*Merged with Cardinal Logistics in February 2013*	Private	$1 046	$1 105	5.60%
18	Saia	LTL	*Saia Motor Freight Line*	Public	$1 030	$1 099	6.60%
19	Kenan Advantage Group	TL	*Acquired Jack B. Kelley in September 2011*	Private	$938	$1 091	10.40%
20	CRST International	TL	*Acquired Specialized Transportation (STI) in July 2011*	Private	$846	$1 051	25.40%
21	Crete Carrier	TL	*Crete Carriers, Shaffer Trucking, Hunt Transportation*	Private	$942	$999	6.10%
22	Roadrunner Transportation	LTL	*Completed 8 acquisitions in 2012. Largest year-over-year growth in the Top 50 Trucking list*	Public	$758	$988	28.60%
23	Knight Transportation	TL	*Dry Van, Refrigerated, Brokerage, Port & Rail Services, Intermodal*	Public	$866	$936	8.10%
24	Southeastern Freight Lines	LTL	*Launched brokerage service in January 2011*	Private	$820	$875	6.60%
25	Averitt Express	LTL	*Revenue growth driven by LTL division*	Private	$739	$739	2.60%

Data source: JOC 2013.

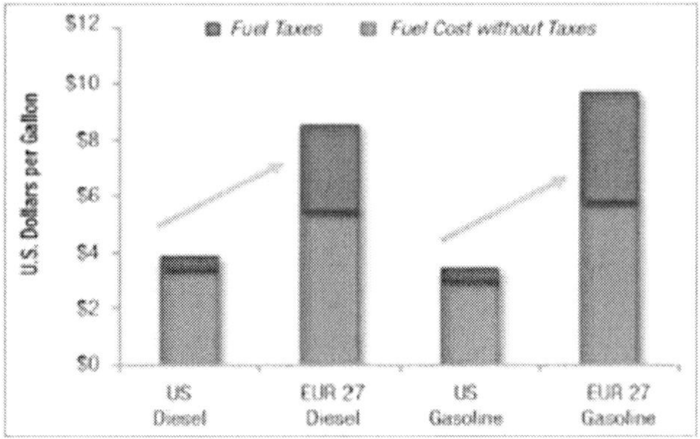

Fuel taxes for the United States include $0.184/gal for gasoline and $0.244/gal for diesel.

Source: U.S. DOE 2010.

Figure 14. Comparison of fuel prices and taxes in the United States and Europe, 2007.

Other transportation segments that could provide business opportunities for the biomass-based diesel industry are railroad; vessel bunkering; and construction, farm, and military equipment. While overall these segments do not consume as much diesel as freight trucks, they could still play a role at local levels. Tables 10-14 in the Appendix illustrate these sectors' diesel consumption by state. The rail network spreads throughout all states but certain areas such as Texas, California, Oklahoma, and Ohio report significantly more diesel consumption by the rail industry. The sector is strong in Texas and California due to a concentration of industrial and agricultural activities, in Oklahoma due to the state's somewhat central location, and in Ohio due to movements between eastern locations and markets in the Midwest. Diesel use by vessel bunkers is naturally high in coastal areas, while highly populated states (namely California and Texas) stand out as the largest consumers of diesel by the construction industry. About half of the country's diesel consumption in farm machinery is concentrated in the Midwest; large states with strong agricultural activities such as California and Texas also report high diesel use by this sector. As noted earlier, the military is not a heavy consumer of diesel; most activities are concentrated in states with large military bases such as Washington, Texas, North Carolina, and California.

Jet fuel is forming as a large and profitable market for the renewable fuels industry. Success in this area could stimulate a significant increase in the production of biofuels and associated feedstock. Successful test flights by commercial and military aircraft over the past few years led to the adoption of specification D7566 by the ASTM International in July 2011, which enables the use of 50% by volume paraffinic renewable fuels with conventional jet fuel. The blending level is restricted to 50% to ensure that the final blend contains a minimum level of aromatic compounds to prevent shrinkage of aged elastomer seals and subsequent fuel leakage. Since then, renewable jet fuel blends have been used by a number of airlines. The German airline Lufthansa began regularly scheduled commercial flights using Neste Oil's NExBTL renewable aviation fuel derived via hydrotreating vegetable oils and animal fats. This made Lufthansa the world's first airline to utilize biofuel in daily flight operations and Neste Oil the first company to provide the biofuel to be used on regularly scheduled commercial flights (Neste Oil 2012). In November 2011, Continental Airlines Flight 1403 from Houston to Chicago was the first U.S. commercial flight to run on a biofuel blend derived partially from genetically modified algae, which was provided by Solazyme. The parent company, United Continental Holdings Inc., plans to buy 20 million gallons per year of algae-derived biofuel made by Solazyme (*Seeking Alpha* 2012). Shortly after, Alaska Airlines and its sister airline Horizon Air started flights using a biofuel blend made from recycled cooking oil. More U.S. airlines are expected to join the effort to fly with renewable jet fuel in the following years.

The U.S. military is a promising client for renewable fuels given the fact that the Department of Defense (DoD) is the single largest consumer of petroleum in the country. The Defense Logistics Agency reports that it purchased about 4.7 billion gallons of fuel in FY 2012, of which 75% represented jet fuel (DLA 2013). Considering the approval of a 50% biofuels blend, this consumption represents a substantial potential demand for biofuels from the U.S. military in the near term. Among all services, the Air Force consumes the most energy and uses more than 2 billion gallons of aviation fuel annually. By 2016, it plans for half of its domestically purchased aviation fuel to be derived from renewable resources. Similarly, the Navy's goal is to use 50% of its energy consumption from alternative sources by 2020. This goal came as a result from successfully testing an F/A-18 fighter jet and helicopters on biofuel blends. The military plays a very strategic role in the development and commercialization of advanced biofuels. It can afford research and testing of alternative fuels to determine their viability more easily than the private sector, which is somewhat limited by concerns about returns on investment. Also, it can purchase large volumes of fuel to create demand, which could lead to increased production and lower prices. Moreover, early adoption of advanced fuel technologies by the DoD provides certainty to investors that there will be a market for these products.

Conclusion

Our examination of the status, opportunities, and feasibility for biomass-based diesel and jet fuel in the United States leads to the following findings:

1) It is technically feasible to produce biomass-derived diesel and jet fuel substitutes in the United States. Many conversion technology options exist. Some are commercially available or in demonstrational stage; others are still in the research and development phase.
2) Biodiesel, consisting of FAME produced from lipids, is currently the predominant form of biomass-based diesel. Production reached a record 1.1 billion gallons in 2011, kept that level in 2012, and is expected to be higher in 2013. Biodiesel blends cannot yet be considered fully "drop-in" fuels because they cannot be transported in all petroleum product pipelines. For pipelines that transport jet fuel, there is a concern that the jet fuel will be contaminated with biodiesel, making it unsuitable for use. Ongoing research aims to determine what, if any, level of FAME can be tolerated in jet fuel.
3) In comparison, the current U.S. renewable diesel and jet fuel production capacity is small, about 225 million gallons per year.

These fuels can be produced from various biomass resources and through several different approaches which all target hydrocarbon products that are similar to petroleum fuels in chemical makeup, and therefore may be considered "drop-in" fuels. It is anticipated that, as "drop-in" fuels, they can be blended with petroleum diesel/jet fuel at high levels, or possibly used in neat form.

- Renewable diesel is produced at commercial scale primarily by hydroisomerization of lipid feedstock. Currently, there are two commercial facilities utilizing this process: Dynamic Fuels (Geismar, Louisiana) and Diamond Green Diesel (Norco, Louisiana).
- A number of processes are under development for production of renewable diesel from biomass-derived sugars (corn, sugarcane, and sorghum, as well as sugars from thermochemical or biochemical depolymerization of cellulose and hemicellulose).
- Processes are also being developed for direct conversion of lignocellulosic biomass by fast pyrolysis, gasification, and other thermochemical means. The first commercial plant is operated by KiOR in Columbus, Mississippi and became operational in early 2013.

4) The costs for producing renewable diesel and jet fuel are not well known and involve a high degree of uncertainty. The process economics for these fuels is highly dependent upon the cost of the feedstock, similar to biodiesel. Additionally, variables such as plant size and co-product credits can have a significant impact on the overall production cost. Hydroisomerization of lipids is performed commercially by Dynamic Fuels and Diamond Green Diesel, and internationally by Neste Oil. The KiOR technology, utilizing pyrolysis, is at an initial commercial scale. Other technology routes are not yet commercial and display a wide range of estimated costs in public sources. As these technology pathways mature and become more widespread, more specific information regarding their economics will be available, which would enable a more detailed analysis and performance comparison.
5) From a feedstock perspective, enough lignocellulosic material is projected to be available in support of the RFS mandate of 21 billion gallons of advanced biofuels. Crop and forest residues alone could yield about 8-24 billion gallons of biomass-based diesel/jet fuel in 2022 (assuming a conversion via fast pyrolysis). This potential could

be larger if the conversion technologies achieve higher yields and if additional feedstock, such as dedicated energy crops, become available. However, there will be competition for lignocellulosic feedstock with the ethanol industry and renewable gasoline producers to meet the RFS mandate. Thus, it is unclear what share the renewable diesel/jet fuel would have in the total biofuels contribution. Ultimately, it will depend on the rate of commercialization of these technologies, selling price, and the transportation market demands.

6) Based on current statistics, and as proven by the biodiesel industry, there is enough lipid feedstock to support the production of 1 billion gallons of biomass-based diesel mandated by the RFS. Today, roughly half of the biodiesel in the United States is produced from soybeans. The remaining portion consists of animal fat, used cooking oil, canola, and some other minor feedstocks. While soybean production is projected to grow in coming years, the biodiesel industry hopes to achieve higher output through advanced technologies for increasing oil supply and production of new feedstock. If algal oil becomes commercially available, as projected within the next 5-10 years, it would greatly benefit both biodiesel and renewable diesel/jet fuel industries. Given the right resources, algal oil productivity can be quite high. Algae are a potential aquatic oil crop, but may also yield carbohydrates that can be converted to sugar.
7) Demand for diesel and jet fuel in the United States is projected to grow. As easily recoverable crude oil resources are diminishing and as their prices rise, more substitutes are expected to enter the market.
   - Among the diesel consumers in the country, freight trucking has the largest share. The number of light-duty vehicles using diesel is projected to increase, the rate of which will depend on the market penetration of other alternatives such as hybrid and electric vehicles.
   - Jet fuel is forming as a large and profitable market for the renewable fuels industry. Success in this area could stimulate a significant increase in the production of biofuels and associated feedstock. It is expected that jet fuel consumption by commercial carriers will continue to grow over the next years, whereas jet fuel consumption by the military will remain flat.
8) For biomass-based diesel and jet fuel to be successful among the trucking and aviation companies, they must be cost-competitive with petroleum-based fuels. It is uncertain what the future holds for these

substitutes, but it is expected that the next several years, as more facilities come online, will answer many questions about the economic viability of these technologies. Much will depend on the rate of recovery of U.S. and world economies, oil prices, carbon market, and political climate.

REFERENCES

"How the New MPG Standards Will Affect Drivers, Automakers, Car Dealerships & More." *Time Magazine*. (2012). August 30 2012, http://business.time.com/2012/08/30/how-the-new-mpg-standards-will-affect-driversautomakers-car-dealerships-more/.

"Terrabon achieves production milestone at Texas demo facility." *Ethanol Producer Magazine*. (2011). January 27, 2011. http://www.ethanol

Advanced Biofuels & Biobased Materials Project Database. *Biofuels Digest.* (2012). Release Q3 2012: http://bit.ly/MP7MHv.

Agusdinata, D.; Zhao, F.; Lleleji, K. (2011). "Life Cycle Assessment of Potential Biojet Fuel Production in the United States." *Environmental Science and Technology* (45); pp. 9133-9143: http://pubs.acs.org/doi /abs/10.1021/es202148g.

American Trucking Association (ATA). (2012). "Renewable Diesel and Biodiesel." Accessed December 2012: http://www.trucking.org/advissues /energy

Amyris. "Breakthrough Science." Accessed July 2012, http://www.amyris biotech.com/Innovation/155/BreakthroughScience.

Anex, R. P.; Aden, A.; Kazi, F. K.; Fortman, J.; Swanson, R. M.; Wright, M. M.; Satrio, J. A.; Brown, R. C.; Daugaard, D. E.; Platon, A.; Kothandaraman, G.; Hsu, D. D.; Dutta, A. (2010). "Techno-Economic Comparison of Biomass-to-Transportation Fuels Via Pyrolysis, Gasification, and Biochemical Pathways." *Fuel.* 89 (1); pp. S29-S35: http://www.nrel.gov/docs/fy11osti/47764.pdf.

Biddy, M.; Jones, S. (2013). " Catalytic Upgrading of Sugars to Hydrocarbons Technology Pathway." Technical Report NREL/TP-5100-58055, PNNL-22319: http://www.nrel.gov/docs/fy13osti/58055.pdf.

Brown, T.R.; Wright, M.M.; Brown, R.C. (2011). "Estimating profitability of two biochar production scenarios: slow pyrolysis *vs* fast pyrolysis." *Biofuels, Bioproducts and Biorefining* (5); pp.54–68: http://onlinelibrary. wiley.com/doi/10.1002/bbb.254/pdf.

Cobalt. "BIOBUTANOL — Versatile Chemical Serving Large Markets". Accessed March 2013: http://www.cobalttech.com/.

Coskata Inc. "The Coskata Process." Accessed January 2011. http://coskata.com/process/.

Davis, R. (2009). "Techno-economic analysis of current technology for Fischer-Tropsch fuels production," NREL Technical Memorandum. National Bioenergy Center. August 14, 2009.

Davis, R.; Aden, A.; Pienkos, P. (2011). "Techno-economic analysis of autotrophic microalgae for fuel production," *Applied Energy* (88); pp. 3524–3531: http://www.sciencedirect.com/science

Davis, R.; Fishman, D.; Frank, E.; Wigmosta, M. (2012). "Renewable Diesel from Algal Lipids: An Integrated Baseline for Cost, Emissions, and Resource Potential from a Harmonized Model." Technical Report ANL/ESD/12-4, NREL/TP-5100-55431, PNNL21437, http://www.nrel.gov/docs/fy12osti/55431.pdf.

Defense Logistics Agency (DLA). (2012). "Fact Book Fiscal Year 2012". http://www.energy 2.pdf.

Dynamic Fuels. (2013). Dynamic Fuels, LLC website. Accessed November 2013. http://www.dynamicfuelsllc.com/about.aspx.

EIA. (2011). "Annual Energy Outlook 2011," April 2011, http://www.eia.gov/forecasts/aeo/.

EIA. (2013b). "Distillate Fuel Oil and Kerosene Sales by End Use," November 2013, http://www.eia.gov/dnav/pet/pet_cons_821use_dcu_nus_a.htm.

EIA. (2013c). "Product Supplied," September 2013, http://www.eia.gov/dnav/pet/pet_cons_psup_dc_nus_mbbl_a.htm.

EIA. (2013d). "State Energy Data System – Jet Fuel Consumption, Price, and Expenditure Estimates, 2012," September 2013, http://www.eia.gov/state fuel/html/fuel jf.html.

EIA. (2013e). "U.S. On-Highway Diesel Fuel Prices," October 2013, http://www.eia.gov/petroleum

Energy Information Administration (EIA). (2013a). "Monthly Energy Review – Biodiesel Overview," October 2013, http://www.eia.gov/totalenergy/data/monthly/#renewable.

Environmental Protection Agency (EPA). (2010a). "Regulation of Fuels and Fuel Additives: 2011 Renewable Fuel Standards," July 2010, http://www.gpo.gov/fdsys/pkg/FR-2010-07-20/pdf/2010-17281.pdf.

EPA (2009). "Renewable Fuel Standard Program (RFS2): Notice of Proposed Rulemaking", http://www.epa.gov/otaq/renewablefuels/rfs2 1-5.pdf.

EPA. (2010b). "Renewable Fuel Standard Program (RFS2)–2010 and Beyond," April 2010, presentation to the Union League Club, http://www.mcguirewoods.com/media. pdf.

EPA. (2013a). "RFS2 EMTS Informational Data," http://www.epa.gov/otaq /fuels/rfsdata/index.htm.

EPA. (2013b). "EPA Finalizes 2013 Renewable Fuels Standards," August 2013, http://www.epa.gov/otaq/fuels/renewablefuels/documents/420f1 3042.pdf.

EPA. (2013c). "EPA Proposes 2014 Renewable Fuels Standards, 2015 Biomass-Based Diesel Volume". November 2013. http://www.epa.gov /otaq/fuels/renewablefuels/documents/420f13048.pdf.

European Biofuels Technology Platform (EBTP). (2009). "Synthetic Biology, 'Modified Metabolism' and Plant Bioetchnology for Production of Advanced Biofuel Molecules and Improved Feedstocks,"Accessed December 2012, http://www.biofuelstp.eu/biotechnology

FAPRI – Iowa State University. (2012). "2012 World Agricultural Outlook," http://www.fapri.iastate.edu/outlook/2012/.

Food and Agricultural Policy Research Institute at the University of Missouri-Columbia (FAPRI-MU). (2009). "Renewable Identification Number Markets: Draft Baseline Table," http://www.fapri.missouri.edu/outreach f.

Gevo. (2011). "Renewable Solutions." http://gevo.com/wpcontent/uploads /2011/05/GEVO-wp-iso-ftf.pdf

Greentech Media. (2010). "LS9's Genetic Breakthrough: Will it Produce Biofuels as Scale?" http://www.greentechmedia.com/articles/read/LS9s-Biofuel-ProductionBreakthrough/.

Hart Energy Consulting. (2009). "Global Biofuels Outlook 2010-2020", http://www.hartenergy.com/Downstream/Research-And-Consulting/Global-BiofuelsOutlook-2010-2020/.

Jegannathan, K.; Eng-Seng, C.; Ravindra, P. (2011) "Economic assessment of biodiesel production: Comparison of alkali and biocatalyst processes." *Renewable and Sustainable Energy Review,* (15); pp. 745–751. http://www.sciencedirect.com/science

Journal of Commerce (JOC). (2013). "Top 50 Trucking Companies". Accessed September 2013: http://www.joc.com/trucking-logistics/

Kalnes, T.; Marker, T.; Shonnard, D.R. (2007). Article A48, *International Journal of Chemical Reactor Engineering,* (5).

KiOR Inc. "Production Facilities". Accessed March 2013. http://www.kior.com/content/?s=6&s2=56&p=56&t=Production-Facilities

Marker, T.; Felix, L.; Linck, M.; Ortaz Toral, P.; Wangerow, J.; Kraus, L.; McLeod, C.; DelPaggio, A.; Tan, E.; Gephart, J.; Gromov, D.; Roberts, M., Purtle, I., Starr, J., Hahn, J., Dorrington, P., Stevens, J., Shonnard, D., and Maleche, E. (2012) "Biomass to Gasoline and Diesel Using Integrated Hydropyrolysis and Hydroconversion." Extended Abstract, AICHE. Gas Technology Institute (GTI) Technical Report, DOI 10.2172/1059031. http://www.osti.gov/bridge/product.biblio.jsp?osti_id=1059031

Milbrandt, A. (2005). "Geographic Perspective on the Current Biomass Resource Availability in the United States." Golden, CO: National Renewable Energy Laboratory (NREL). NREL/TP-560-39181; 70 pp: http://www.nrel.gov/docs/fy06osti/39181.pdf.

NABC. (2012b). "Hydrothermal Liquefaction – Results from Stage I Extension". January 5, 2012. http://www.nabcprojects.org/pdfs/hydrothermal_liquefaction_results_stage_i_extension.pdf

NAS. (2009). "Liquid transportation fuels from coal and biomass: technological status, costs, and environmental impacts." Washington DC, 2009 http://sites.nationalacademies.org/xpedio/groups/energysite/documents/webpage/energy 054519.pdf

National Advanced Biofuels Consortium (NABC). (2012a). "A Refiner's Perspective on Advanced Biofuels", Prepared from a talk delivered at the at the January 2012 annual NABC meeting by Rick Weyen of Tesoro. http://www.nabcprojects.org/pdfs/refiner_perspective_advanced_biofuels.pdf

National Energy Technology Laboratory (NETL). (2009). "Affordable, Low-Carbon Diesel Fuel from Domestic Coal and Biomass". DOE/NETL-2009/1349. http://www.netl.doe.gov/energy

Neste Oil. "NExBTL Aviation Fuel". Accessed January 2012, http://www.nesteoil.com/default.asp?path=1,41,11991,12243,17555

Pacific Northwest National Laboratory (PNNL). (2009). "Production of Gasoline and Diesel from Biomass via Fast Pyrolysis, Hydrotreating and Hydrocracking: A Design Case". DOE/PNNL-18284. http://www.pnl.gov/main/publications/external/technical_reports/PNNL-18284.pdf

Pearlson,M., Wollersheim, C., and Hileman, J. (2012) "A techno-economic review of hydroprocessed renewables esters and fatty acids for jet fuel production." Biofuels, Bioproducts, and Biorefineries 7:89-96 (2013); DOI: 10.1002/bbb. http://onlinelibrary.wiley.com/doi/10.1002 /bbb.1378 /pdf

PR Newswire. (2012). "U.S. Clean Diesel Auto Sales Increase 25.6 Percent in 2012", http://www.prnewswire.com/news-releases/us-clean-diesel-auto-sales-increase-256- percent-in-2012-181048881.html

Seeking Alpha. (2012). "The U.S. EPA Approves Solazyme's Fuel For Commercial Use." http://seekingalpha.com/article/778491-the-u-s-epa-approves-solazymes-fuel-forcommercial-use

Smagala, T.G.; Christensen, E.; Christison, K.M.; Mohler, R.E.; Gjersing, E.; McCormick, R.L. (2013). "Hydrocarbon Renewable and Synthetic Diesel Fuel Blendstocks: Composition and Properties." *Energy Fuels* 27 (1); pp. 237-246.

Soybean Review. (2011). "RFS2 Provides Stability for Biodiesel's Future", http://soybeanreview.com/article/rfs2-provides-stability

Syntroleum. (2011). "Syntroleum Announces Record July Production and Second Quarter Results," http://www.syntroleum.com/

Syntroleum. (2013). Syntroleum Corporation 2013 Form 10-K (Annual Report). Filed 03/15/13 for the Period Ending 12/31/12.

Tao, L.; Aden, A. (2009). "The economics of current and future biofuels." *In Vitro Cellular & Developmental Biology – Plant*, (45); pp. 199–217. http://link.springer.com/article/10.1007%2Fs11627-009-9216-8

The City Wire. (2013a). "Dynamic Fuels Plant on Standby, Indefinitely." August 7, 2013, http://www.thecitywire.com/node/28947#.Upzkn8SkqDs.

The City Wire. (2013b). "Tyson's Dynamic Fuels Plant Burns Cash." November 18, 2013, http://www.thecitywire.com/node/30545 #.Upzkic SkqDs.

The National Academy of Sciences (NAS). (2011). Committee on Economic and Environmental Impacts of Increasing Biofuels Production; National Research Council. "Renewable Fuel Standard: Potential Economic and Environmental Effects of U.S. Biofuel Policy," The National Academies Press, pp. 131-134, http://www.nap.edu/catalog.php?record_id=13105

U.S. Census Bureau. (2010). "M311K - Fats and Oils: Production, Consumption, and Stocks," http://www.census.gov/manufacturing

U.S. Department of Agriculture (USDA). (2012). "Oil Crops Yearbook 2011" March 2012, http://usda.mannlib.cornell.edu/MannUsda/viewDocument Info.do;jsessionid=53C6FDE58607C6FDCDBF42000CF26338? documentID=1290.

U.S. Department of Energy (USDOE). (2010). "Diesel Power: Clean Vehicles for Tomorrow," http://www1.eere.energy

USDA. (2013). Economic Research Service. "Oil Crops Yearbook 2012". March 2013, http://www.ers.usda.gov/data-products/oil RDsqg

USDOE. (2011). *U.S. Billion-Ton Update: Biomass Supply for a Bioenergy and Bioproducts Industry*. R.D. Perlack and B.J. Stokes (Leads), ORNL/TM-2011/224. Oak Ridge National Laboratory, Oak Ridge, TN, http://www1.eere.energy

Wigmosta, M.; Coleman, A.; Skaggs, R.; Huessemann, M.; Lane, L. (2011). "National Microalgae Biofuel Production Potential and Resource Demand", *Water Resources Research* (47): http://www.agu.org /journals/wr/wr1104/2010WR009966/2010WR009966.pdf

Wright, M.; Brown, R.; Boateng, A. (2008). "Distributed processing of biomass to bio-oil for subsequent production of Fischer-Tropsch liquids."*Biofuels, Bioproducts, and Biorefining* (2); pp. 229–238. http://onlinelibrary.wiley.com/doi/10.1002/bbb.73/pdf

Wright, M.; Daugaard, D.; Satrio, J.; Brown, R. (2009). "Techno-economic analysis of biomass fast pyrolysis to transportation fuels." Proceedings of the 238th National Meeting and Exposition of the American Chemical Society.

Wright, M.; Daugaard, D.; Satrio, J.; Brown, R. (2010). "Techno-economic analysis of biomass fast pyrolysis to transportation fuels". *Fuel* (89); pp. S2–S10. http://www.sciencedirect.com/science

Yusuf, N.; Kamarudin, S.; Yaakub, Z. (2011). "Overview on the current trends in biodiesel production." *Energy Conversion and Management,* (52); pp. 2741–2751.

ZeaChem. (2011). "Technology Overview." Accessed January 2011, http://www.zeachem.com/technology

APPENDIX

Table 8. Distillate Fuel Oil and Kerosene Sales by End Use

	2007	2008	2009	2010	2011	2012
Residential						
Distillate Fuel Oil	5,141,642	5,568,066	4,103,881	3,930,517	3,625,747	3,473,310
No. 1	81,992	83,379	83,828	65,325	67,530	53,219
No. 2	5,059,651	5,484,687	4,020,053	3,865,192	3,558,216	3,420,091
Kerosene	326,320	167,606	266,136	216,647	137,232	67,316
Commercial						
Distillate Fuel Oil	2,718,674	2,650,895	2,785,246	2,738,384	2,715,330	2,557,543
No. 1 Distillate	64,089	89,322	83,323	70,671	63,062	60,566
No. 2 Distillate	2,475,571	2,630,528	2,559,629	2,545,295	2,524,697	2,384,977
No. 2 Fuel Oil	1,217,835	1,137,514	890,809	885,331	771,343	621,683
Ultra Low Sulfur Diesel	731,582	1,026,130	1,331,244	1,418,516	1,630,000	1,662,447
Low Sulfur Diesel	245,223	310,052	207,287	175,838	85,497	83,036
High Sulfur Diesel	280,932	162,132	130,288	65,520	35,918	17,881
No. 4 Fuel Oil	179,014	140,044	142,295	122,420	126,655	111,970
Residual Fuel Oil	481,368	403,977	415,187	366,363	316,713	226,140
Kerosene	57,360	32,257	31,291	35,176	23,646	8,909
Industrial						
Distillate Fuel Oil	2,484,906	2,593,760	2,159,428	2,045,164	2,179,063	2,325,583
No. 1 Distillate	37,787	35,510	26,931	32,532	63,588	39,364
No. 2 Distillate	2,371,370	2,534,715	2,054,033	1,948,268	2,050,831	2,215,033
No. 2 Fuel Oil	324,891	310,686	154,750	122,870	143,117	125,447
Low Sulfur Diesel	1,319,312	1,833,331	1,603,685	1,671,967	1,849,222	2,047,177
High Sulfur Diesel	727,167	390,699	296,278	153,432	93,492	42,410
No. 4 Fuel Oil	87,749	23,626	77,864	64,363	60,534	71,196
Residual Fuel Oil	1,187,319	1,043,883	726,210	667,672	772,676	484,957
Kerosene	66,372	22,445	25,536	45,145	22,557	12,920
Farm						
Distillate Fuel Oil	3,202,847	3,744,536	2,660,024	2,928,175	2,942,436	3,031,878
Diesel	3,147,431	3,698,265	2,620,378	2,895,104	2,912,647	3,007,499
Other Distillate	55,416	46,271	39,647	33,071	29,780	24,370
Kerosene	9,531	4,693	6,414	6,763	3,410	1,712
Electric Power						
Distillate Fuel Oil	669,061	615,626	681,386	648,144	606,693	461,694
Residual Fuel Oil	2,648,574	1,654,584	1,111,384	1,030,682	871,565	467,726
Oil Company						
Distillate Fuel Oil	774,904	1,966,608	760,077	951,322	1,381,127	1,710,513
Residual Fuel Oil	43,977	67,614	75,166	20,783	18,759	17,011
Total Transportation (Railroad, Vessel Bunkering, On Highway)						
Distillate Fuel Oil	45,298,231	42,741,511	38,819,930	40,560,191	41,414,854	41,229,545
Residual Fuel Oil	6,326,931	5,257,810	4,589,049	5,142,573	4,560,070	4,019,580
Railroad						
Distillate Fuel Oil	3,634,512	3,229,629	2,758,140	2,874,041	3,121,150	3,118,150
Vessel Bunkering						
Distillate Fuel Oil	1,573,981	1,883,427	1,912,984	2,087,834	2,133,385	1,768,324
Residual Fuel Oil	6,326,931	5,257,810	4,589,049	5,142,573	4,560,070	4,819,588
On Highway						
Distillate Fuel Oil	39,601,744	37,628,464	34,147,806	35,582,625	36,160,368	36,343,072
Military						
Distillate Fuel Oil	363,145	279,575	243,728	243,242	246,243	142,695
Diesel	202,873	264,637	213,613	216,218	223,841	126,268
Other Distillate	160,272	16,438	30,085	28,025	22,463	16,429
Residual Fuel Oil	17,719	9,250	14,609	9,851	14,653	10,324
Off Highway						
Distillate Fuel Oil	2,512,394	2,605,660	1,985,592	2,148,677	2,070,260	2,088,157
Construction	2,206,889	2,248,506	1,749,599	1,809,806	1,810,394	1,852,241
Non-Construction	306,406	357,155	235,993	338,870	259,866	235,916
All Other						
Distillate Fuel Oil	0	0	0	0	0	0
Residual Fuel Oil	2,606	3,749	6,583	5,860	2,664	1,418
Kerosene	1,520	1,036	633	2,297	899	245

In thousand gallons. Data source: EIA 2013b.

Table 9. No. 2 Diesel Sales for On-Highway Use

	2007	2008	2009	2010	2011	2012
U.S.	39,801,744	37,528,464	34,147,806	35,582,625	36,160,308	36,343,072
East Coast (PADD 1)	11,739,455	10,763,333	9,929,426	10,367,337	10,332,863	10,257,620
New England (PADD 1A)	1,123,563	1,062,422	1,044,171	1,052,933	1,064,679	1,063,943
Connecticut	303,570	292,688	266,121	267,948	271,070	266,474
Maine	181,010	180,284	176,004	178,774	176,241	173,717
Massachusetts	402,629	375,414	399,575	394,967	410,375	418,613
New Hampshire	97,695	97,837	92,882	93,381	91,778	91,143
Rhode Island	73,882	55,826	55,684	59,220	57,035	54,666
Vermont	64,777	60,373	53,905	58,643	58,180	59,330
Central Atlantic (PADD 1B)	4,232,345	3,956,845	3,629,109	3,796,971	3,883,499	3,898,410
Delaware	66,271	59,569	55,610	55,342	55,904	56,530
District of Columbia	10,710	15,887	11,944	11,975	8,742	8,535
Maryland	573,321	528,862	515,315	536,150	501,897	503,999
New Jersey	988,732	902,201	679,361	793,752	888,154	784,867
New York	1,118,505	1,120,424	1,064,997	1,083,536	1,082,247	1,100,897
Pennsylvania	1,474,806	1,329,902	1,301,882	1,316,216	1,346,555	1,443,582
Lower Atlantic (PADD 1C)	6,383,547	5,744,066	5,256,146	5,517,433	5,384,685	5,295,267
Florida	1,670,536	1,478,514	1,322,703	1,340,494	1,329,312	1,340,337
Georgia	1,508,118	1,325,065	1,197,220	1,260,672	1,187,538	1,121,051
North Carolina	1,076,721	968,520	886,949	940,838	954,190	920,147
South Carolina	722,868	667,949	641,944	715,795	715,136	647,447
Virginia	1,111,050	1,017,285	935,552	969,057	917,431	971,673
West Virginia	294,254	286,733	271,778	290,577	281,078	294,612
Midwest (PADD 2)	12,627,117	12,123,237	10,905,027	11,509,060	11,784,798	11,886,665
Illinois	1,516,293	1,436,180	1,364,571	1,340,323	1,454,548	1,387,514
Indiana	1,356,498	1,302,135	1,109,103	1,207,711	1,225,199	1,247,331
Iowa	648,924	643,923	601,324	633,299	641,795	643,505
Kansas	489,575	480,947	441,256	476,191	467,779	474,517
Kentucky	875,135	821,452	742,976	785,240	772,677	767,231
Michigan	906,538	838,960	780,954	819,342	820,526	837,612
Minnesota	673,758	663,548	584,807	610,447	639,520	653,051
Missouri	1,087,199	985,190	944,991	977,111	970,936	957,624
Nebraska	434,093	406,238	386,874	425,442	421,847	418,302
North Dakota	177,467	193,615	194,777	237,443	318,231	376,496
Ohio	1,597,741	1,494,384	1,323,991	1,437,404	1,455,547	1,448,059
Oklahoma	840,366	906,265	729,736	741,369	789,449	814,298
South Dakota	202,607	198,713	197,232	213,298	207,311	227,663
Tennessee	1,061,941	1,002,320	826,077	891,637	895,097	882,475
Wisconsin	758,982	749,367	676,358	712,803	704,336	750,987
Gulf Coast (PADD 3)	7,615,389	7,338,939	6,641,445	6,947,707	7,234,843	7,428,440
Alabama	848,402	736,640	657,070	711,371	717,466	705,904
Arkansas	673,079	627,835	592,469	618,731	605,275	605,758
Louisiana	703,016	690,551	694,073	732,017	749,400	667,605
Mississippi	625,746	623,455	553,866	562,673	552,909	578,156
New Mexico	531,013	473,827	432,794	472,924	495,600	495,026
Texas	4,234,133	4,186,631	3,711,173	3,849,991	4,114,193	4,375,991
Rocky Mountain (PADD 4)	1,999,274	1,891,875	1,693,227	1,768,306	1,800,630	1,854,474
Colorado	588,910	566,484	510,211	523,056	501,361	517,545
Idaho	275,706	243,868	225,624	254,417	254,007	255,815
Montana	263,696	250,069	235,067	243,660	251,686	259,418
Utah	480,993	446,071	399,528	401,810	474,895	462,240
Wyoming	389,969	385,383	322,797	345,363	318,681	359,456
West Coast (PADD 5)	5,820,509	5,411,080	4,978,681	4,990,215	5,007,174	4,915,873
Alaska	174,112	171,530	202,102	166,599	169,158	111,113
Arizona	865,536	792,698	739,863	741,588	757,789	742,679
California	3,091,491	2,838,723	2,591,988	2,602,646	2,633,352	2,603,546
Hawaii	52,692	56,394	46,847	50,187	45,792	47,897
Nevada	378,182	329,292	302,145	298,895	288,458	289,875
Oregon	560,598	534,041	498,335	513,521	513,970	507,784
Washington	697,898	688,302	597,401	616,779	598,655	612,979

In thousand gallons. Data source: EIA 2013b.

Table 10. Diesel Sales for Farm Use

	2007	2008	2009	2010	2011	2012
U.S.	3,147,431	3,698,265	2,620,378	2,895,104	2,912,647	3,007,499
East Coast (PADD 1)	358,009	377,605	326,068	449,575	368,704	374,504
New England (PADD 1A)	18,818	20,214	11,470	10,956	14,953	13,938
Connecticut	1,798	2,160	987	1,091	1,515	1,880
Maine	6,509	7,547	2,350	3,713	4,335	4,399
Massachusetts	3,273	1,240	1,223	831	1,195	1,831
New Hampshire	2,229	2,622	2,083	772	1,263	1,295
Rhode Island	40	103	20	16	23	34
Vermont	4,970	6,543	4,808	4,532	6,622	4,499
Central Atlantic (PADD 1B)	96,794	113,257	90,495	99,578	104,693	99,395
Delaware	5,839	4,762	5,903	6,821	8,541	6,767
District of Columbia	--	0	0	0	0	0
Maryland	13,677	17,618	8,020	10,720	11,420	11,292
New Jersey	2,030	2,726	5,951	7,343	6,115	6,413
New York	39,923	43,886	30,246	20,092	29,844	30,636
Pennsylvania	35,325	44,266	40,376	54,603	48,773	44,287
Lower Atlantic (PADD 1C)	242,397	244,133	224,103	339,041	249,059	261,171
Florida	69,001	87,131	86,642	204,791	109,128	103,325
Georgia	68,172	47,824	69,053	62,281	63,725	79,470
North Carolina	49,930	50,645	41,443	39,454	38,571	42,955
South Carolina	20,374	20,058	8,176	9,379	12,217	10,832
Virginia	33,347	36,810	17,192	20,263	23,370	22,452
West Virginia	1,573	1,667	1,596	2,874	2,048	2,137
Midwest (PADD 2)	1,596,524	1,844,300	1,384,186	1,502,450	1,510,797	1,599,681
Illinois	175,985	210,548	119,015	142,946	152,491	151,626
Indiana	136,331	130,194	92,486	85,257	116,498	126,061
Iowa	155,346	201,338	178,089	200,458	191,678	202,588
Kansas	148,827	189,901	138,644	156,717	135,899	135,168
Kentucky	24,308	30,569	20,187	18,594	20,430	18,582
Michigan	47,329	53,828	49,200	56,779	60,122	50,302
Minnesota	114,924	131,577	153,716	166,789	158,840	157,524
Missouri	138,517	123,377	88,445	91,575	92,162	101,940
Nebraska	199,851	233,656	149,308	142,888	142,327	199,787
North Dakota	100,554	136,067	96,106	119,470	134,935	126,979
Ohio	107,688	130,449	95,777	106,136	87,699	106,475
Oklahoma	57,403	56,885	36,066	43,851	26,505	33,172
South Dakota	59,835	67,897	63,165	56,899	67,198	64,939
Tennessee	36,162	31,532	20,400	23,276	21,851	23,379
Wisconsin	93,464	116,481	83,583	90,816	102,163	101,158
Gulf Coast (PADD 3)	604,338	735,740	381,868	434,193	462,577	468,982
Alabama	32,659	44,138	17,882	19,881	24,518	24,503
Arkansas	212,255	222,990	68,641	92,965	83,323	87,118
Louisiana	51,507	58,838	41,219	42,255	48,056	45,753
Mississippi	50,123	41,385	41,080	57,087	52,559	81,878
New Mexico	15,472	20,028	11,277	14,821	10,950	12,816
Texas	242,322	348,361	201,769	207,183	243,170	216,915
Rocky Mountain (PADD 4)	207,302	227,081	203,467	167,704	175,686	159,171
Colorado	43,695	58,955	40,518	39,603	41,532	32,759
Idaho	54,577	55,455	53,759	66,442	70,724	61,287
Montana	94,618	96,116	93,447	44,921	44,840	47,207
Utah	5,885	5,110	6,786	5,831	6,794	7,031
Wyoming	8,526	11,445	8,957	10,907	11,797	10,886
West Coast (PADD 5)	381,259	513,539	324,788	341,182	394,882	405,161
Alaska	37	65	108	109	126	177
Arizona	19,184	35,197	35,893	34,514	39,760	35,964
California	268,177	349,107	187,956	205,254	244,358	258,004
Hawaii	4,102	3,570	4,175	3,493	3,483	3,074
Nevada	3,008	2,901	5,322	6,351	6,444	5,102
Oregon	26,982	30,579	33,964	33,174	46,209	47,498
Washington	59,769	92,119	57,371	58,287	54,502	55,342

In thousand gallons. Data source: EIA 2013b.

Table 11. Sales of Distillate Fuel Oil for Railroad Use

	2007	2008	2009	2010	2011	2012
U.S.	3,634,512	3,229,625	2,759,140	2,974,641	3,121,150	3,118,150
East Coast (PADD 1)	580,632	500,071	459,324	482,929	514,418	492,156
New England (PADD 1A)	69,282	47,582	43,763	53,930	51,126	33,306
Connecticut	4,450	3,219	2,219	2,006	2,006	5,195
Maine	126	1,694	7,252	8,284	6,818	5,970
Massachusetts	63,896	40,378	24,852	33,130	32,647	12,307
New Hampshire	119	126	697	86	124	116
Rhode Island	13	72	4	24	3	133
Vermont	678	2,092	8,740	10,400	9,528	9,586
Central Atlantic (PADD 1B)	210,461	177,750	152,309	196,570	233,005	204,527
Delaware	1,404	1,120	1,096	879	126	149
District of Columbia	0	0	0	1,229	6,392	6,770
Maryland	11,546	3,214	17,035	34,717	36,283	20,384
New Jersey	15,616	15,055	8,071	1,778	1,660	1,325
New York	63,226	44,510	35,307	33,709	42,254	35,237
Pennsylvania	118,670	113,851	90,800	124,258	146,291	140,663
Lower Atlantic (PADD 1C)	300,889	274,739	263,252	232,429	230,287	254,322
Florida	74,409	64,963	33,651	42,353	46,461	66,711
Georgia	78,927	69,710	62,072	63,770	71,374	63,902
North Carolina	47,855	29,022	89,823	62,103	32,158	41,501
South Carolina	11,321	16,023	3,602	3,051	3,973	3,983
Virginia	72,611	79,606	63,960	49,503	63,611	67,769
West Virginia	15,766	15,416	10,143	11,650	12,711	10,456
Midwest (PADD 2)	1,561,277	1,420,396	1,144,926	1,223,206	1,215,528	1,195,263
Illinois	40,116	51,287	55,322	72,188	58,526	63,808
Indiana	65,820	51,232	37,773	50,736	63,437	68,061
Iowa	58,640	62,458	40,494	41,663	36,136	30,156
Kansas	92,323	129,141	147,106	78,143	80,404	99,475
Kentucky	170,042	94,124	48,002	42,101	67,347	61,840
Michigan	49,528	41,887	25,920	18,376	10,330	13,352
Minnesota	123,390	78,651	39,188	47,567	61,340	92,275
Missouri	27,467	13,281	19,765	36,396	51,179	44,914
Nebraska	12,732	27,507	75,064	214,176	181,421	166,060
North Dakota	124,832	58,667	12,849	8,983	9,839	43,907
Ohio	333,069	316,926	206,134	179,048	203,135	175,258
Oklahoma	348,832	395,252	352,301	349,077	313,806	245,376
South Dakota	8,572	10,024	5,730	5,860	7,182	10,826
Tennessee	76,692	41,676	53,391	54,332	58,125	52,522
Wisconsin	29,222	48,285	25,886	24,559	13,321	27,434
Gulf Coast (PADD 3)	699,882	631,796	542,036	573,037	694,053	729,109
Alabama	59,852	42,588	44,546	42,465	97,177	125,439
Arkansas	20,237	27,693	25,148	18,302	26,907	43,494
Louisiana	43,862	32,201	18,345	25,425	32,515	28,110
Mississippi	46,730	31,617	24,727	17,936	37,741	29,848
New Mexico	6,152	2,092	245	1,780	1,707	19,123
Texas	523,049	495,604	429,026	467,128	498,006	483,096
Rocky Mountain (PADD 4)	262,644	222,054	212,571	228,200	245,446	214,160
Colorado	4,014	5,422	47,830	66,510	71,365	77,038
Idaho	21,070	14,622	9,678	31,307	30,448	25,068
Montana	107,710	94,818	68,520	58,543	65,919	41,901
Utah	44,721	24,643	21,178	24,774	33,371	24,216
Wyoming	85,129	82,549	65,365	47,065	44,344	45,938
West Coast (PADD 5)	530,077	455,308	400,283	467,270	451,704	487,461
Alaska	6,419	6,120	5,899	5,399	5,754	5,564
Arizona	11,940	7,230	8,200	10,566	9,698	20,624
California	317,292	261,225	219,854	252,057	255,313	258,354
Hawaii	4	6	5	37	4	4
Nevada	6,874	7,101	11,594	7,446	8	44
Oregon	82,369	94,925	91,305	104,445	87,470	114,507
Washington	105,180	78,701	63,425	87,321	93,458	88,364

In thousand gallons. Data source: EIA 2013b.

Table 12. Sales of Distillate Fuel Oil for Off-Highway Use

	2007	2008	2009	2010	2011	2012
U.S.	2,512,394	2,605,660	1,985,592	2,148,677	2,070,260	2,088,157
East Coast (PADD 1)	833,519	883,356	605,884	615,812	634,470	621,261
New England (PADD 1A)	92,754	113,790	81,453	102,263	102,751	75,212
Connecticut	21,159	19,948	14,456	16,124	16,435	10,683
Maine	12,193	15,262	14,483	15,495	16,622	18,373
Massachusetts	39,016	56,006	27,388	43,133	43,432	19,129
New Hampshire	11,495	14,814	8,898	12,689	11,421	10,558
Rhode Island	4,540	2,129	5,652	3,821	3,725	3,057
Vermont	4,352	5,632	10,576	11,000	11,116	13,413
Central Atlantic (PADD 1B)	226,685	252,027	186,785	187,163	213,795	208,407
Delaware	3,149	3,210	2,578	2,201	2,306	1,812
District of Columbia	1,988	1,223	1,043	357	920	892
Maryland	35,368	37,065	15,198	14,402	23,521	21,838
New Jersey	65,404	65,491	48,769	47,895	62,894	59,809
New York	51,681	48,869	37,033	34,272	38,054	35,089
Pennsylvania	69,096	96,169	82,164	88,035	86,101	88,967
Lower Atlantic (PADD 1C)	514,080	517,539	337,646	326,386	317,924	337,641
Florida	124,123	135,397	112,263	110,675	104,005	109,119
Georgia	124,009	111,648	75,856	79,376	76,496	81,370
North Carolina	53,366	52,272	43,755	53,910	51,483	42,119
South Carolina	51,781	57,191	36,537	35,998	29,800	41,179
Virginia	138,601	139,537	59,539	38,195	46,783	56,139
West Virginia	22,201	21,494	9,695	8,231	9,357	7,715
Midwest (PADD 2)	686,931	716,068	544,191	557,179	524,627	554,255
Illinois	110,067	108,788	60,997	59,541	56,167	66,746
Indiana	33,460	50,166	52,110	41,855	49,234	49,992
Iowa	24,765	28,979	29,836	30,989	26,475	30,634
Kansas	21,572	17,213	28,999	30,359	25,422	19,184
Kentucky	26,757	34,351	31,866	27,906	25,590	21,723
Michigan	47,782	52,039	35,267	34,737	35,147	33,514
Minnesota	57,964	86,484	45,198	54,211	62,268	55,294
Missouri	68,262	64,660	50,840	51,485	41,060	31,435
Nebraska	44,362	11,846	26,223	18,344	14,586	19,173
North Dakota	11,741	11,056	25,623	27,802	25,811	44,192
Ohio	68,156	74,876	56,113	56,694	49,976	60,223
Oklahoma	44,661	45,228	25,704	24,332	24,912	27,796
South Dakota	10,282	9,260	9,622	7,239	15,484	10,211
Tennessee	65,938	68,110	31,117	43,251	37,539	43,024
Wisconsin	51,163	53,012	34,676	38,436	34,954	41,115
Gulf Coast (PADD 3)	480,332	486,148	360,795	382,691	370,941	417,101
Alabama	62,743	80,762	61,501	58,051	58,172	69,907
Arkansas	48,044	30,752	23,210	29,752	22,773	22,622
Louisiana	76,131	82,523	48,510	68,889	44,130	42,719
Mississippi	59,253	53,833	17,862	18,404	20,889	26,199
New Mexico	3,235	16,705	5,729	24,907	24,865	29,454
Texas	230,926	221,572	203,983	182,689	200,112	226,200
Rocky Mountain (PADD 4)	149,317	138,282	139,107	121,101	125,195	113,299
Colorado	48,550	47,759	52,113	48,837	56,745	45,876
Idaho	23,362	18,349	14,912	11,780	11,916	10,317
Montana	22,325	28,642	20,985	19,520	14,700	14,814
Utah	35,219	29,871	26,922	19,952	26,754	20,749
Wyoming	19,860	13,661	24,176	21,012	15,080	21,543
West Coast (PADD 5)	362,295	381,807	335,615	471,893	415,027	382,242
Alaska	16,599	22,832	14,334	14,738	10,126	13,962
Arizona	66,808	76,168	49,894	60,142	65,414	66,025
California	135,327	127,748	146,296	195,299	208,632	183,553
Hawaii	9,414	8,570	9,806	7,839	7,920	9,578
Nevada	29,733	29,091	43,385	124,123	44,879	37,652
Oregon	22,925	33,848	34,306	30,108	34,562	36,028
Washington	81,488	83,550	37,593	39,644	43,493	35,443

In thousand gallons. Data source: EIA 2013b.

Table 13. Distillate Fuel Oil Sales for Vessel Bunkering Use

	2007	2008	2009	2010	2011	2012
U.S.	1,923,981	1,983,422	1,912,984	2,002,834	2,133,395	1,768,324
East Coast (PADD 1)	466,132	461,533	276,013	259,319	296,947	283,254
New England (PADD 1A)	43,014	69,102	45,147	30,589	32,414	38,891
Connecticut	6,654	5,683	3,914	1,898	1,502	2,838
Maine	8,298	6,815	15,611	4,207	4,128	13,349
Massachusetts	21,336	48,094	19,193	17,529	17,132	13,612
New Hampshire	2,740	2,552	2,327	1,110	1,395	1,815
Rhode Island	3,987	5,958	4,101	5,824	8,257	7,243
Vermont	0	0	0	21	0	35
Central Atlantic (PADD 1B)	147,629	129,789	104,487	67,726	76,446	74,154
Delaware	615	919	582	485	1,658	615
District of Columbia	11	7	5	13	15	17
Maryland	21,380	16,507	8,240	8,335	2,662	7,300
New Jersey	87,549	74,725	68,321	37,276	61,990	55,842
New York	12,339	10,814	8,497	6,869	4,453	7,089
Pennsylvania	25,735	26,816	18,842	14,748	5,669	3,290
Lower Atlantic (PADD 1C)	275,489	262,642	126,379	161,005	188,087	170,209
Florida	145,269	180,514	84,718	118,991	142,198	131,685
Georgia	14,016	10,831	10,765	12,904	12,387	11,300
North Carolina	8,009	8,280	5,957	8,231	5,613	4,177
South Carolina	18,795	15,892	8,094	8,043	5,865	4,966
Virginia	43,971	18,558	16,745	12,731	21,562	18,063
West Virginia	45,429	28,568	99	105	461	18
Midwest (PADD 2)	386,653	428,712	385,719	370,587	389,493	354,341
Illinois	71,805	101,851	85,117	56,575	56,000	59,508
Indiana	29,805	24,226	17,165	19,849	22,319	8,139
Iowa	4,867	4,151	4,523	5,381	5,163	5,609
Kansas	0	0	0	0	0	0
Kentucky	01,516	104,307	102,305	98,294	104,160	96,669
Michigan	8,900	7,585	7,585	17,875	19,825	20,792
Minnesota	7,551	18,987	9,811	7,638	4,111	1,649
Missouri	34,158	42,760	44,967	49,922	42,388	40,858
Nebraska	0	0	0	0	0	0
North Dakota	0	0	0	0	0	0
Ohio	12,122	17,733	24,586	27,668	30,684	23,532
Oklahoma	10,764	10	8	9	6	7
South Dakota	0	0	0	0	0	0
Tennessee	114,703	106,326	88,312	85,786	103,077	95,856
Wisconsin	461	696	1,339	1,591	1,760	1,723
Gulf Coast (PADD 3)	613,864	721,875	827,977	1,010,776	1,026,408	677,343
Alabama	36,334	63,100	61,852	65,017	41,339	25,542
Arkansas	420	49,384	47,557	47,769	55,246	425
Louisiana	351,726	399,347	378,642	481,453	564,650	382,462
Mississippi	93,581	106,799	141,302	93,384	58,285	58,505
New Mexico	0	0	0	0	0	0
Texas	131,803	103,245	198,625	323,153	306,887	210,408
Rocky Mountain (PADD 4)	27	26	19	6	2	17
Colorado	2	2	1	1	2	2
Idaho	25	25	18	5	0	15
Montana	0	0	0	0	0	0
Utah	0	0	0	0	0	0
Wyoming	0	0	0	0	0	0
West Coast (PADD 5)	457,306	371,275	423,256	362,147	420,545	453,369
Alaska	127,614	119,460	104,126	115,129	118,900	136,681
Arizona	1	1	0	1	2	1
California	104,583	88,088	163,438	96,015	115,772	127,038
Hawaii	129,743	58,786	71,496	63,757	88,652	86,663
Nevada	25	22	7	6	6	7
Oregon	20,502	24,003	24,143	22,626	21,728	22,450
Washington	74,838	80,915	60,046	64,613	75,485	80,528

In thousand gallons. Data source: EIA 2013b.

Table 14. Diesel Sales for Military Use

	2007	2008	2009	2010	2011	2012
U.S.	202,873	254,537	213,643	215,218	223,841	126,268
East Coast (PADD 1)	43,728	57,003	47,877	42,842	58,296	35,470
New England (PADD 1A)	8,260	15,028	3,342	3,427	5,656	2,836
Connecticut	1,660	997	207	236	622	501
Maine	3,551	6,915	1,094	2,495	1,300	774
Massachusetts	2,125	3,182	500	343	3,101	271
New Hampshire	815	3,178	1,225	292	143	83
Rhode Island	19	399	35	21	468	1,091
Vermont	90	357	282	40	22	115
Central Atlantic (PADD 1B)	12,921	15,306	10,429	9,111	13,076	8,293
Delaware	99	88	122	75	168	70
District of Columbia	598	291	165	265	693	300
Maryland	3,950	4,772	2,807	3,249	3,646	1,162
New Jersey	3,951	4,562	2,331	2,127	2,408	3,323
New York	3,058	3,656	3,288	2,460	2,394	785
Pennsylvania	1,265	1,939	1,715	936	3,767	2,654
Lower Atlantic (PADD 1C)	22,547	26,669	34,106	30,304	39,564	24,342
Florida	4,444	6,115	4,370	5,481	6,323	6,043
Georgia	2,502	2,936	3,644	3,282	3,357	2,957
North Carolina	3,848	5,265	18,456	16,839	12,133	2,794
South Carolina	6,479	8,859	4,029	1,911	1,436	636
Virginia	4,812	3,460	3,259	2,611	16,167	11,630
West Virginia	462	33	347	179	148	281
Midwest (PADD 2)	7,549	8,934	8,526	7,053	7,483	5,948
Illinois	266	316	329	728	1,415	239
Indiana	12	225	46	121	31	45
Iowa	38	37	50	50	154	43
Kansas	812	614	586	436	341	87
Kentucky	919	944	470	1,110	1,022	467
Michigan	1,184	2,147	889	267	807	1,270
Minnesota	292	454	257	167	225	322
Missouri	1,738	1,939	1,881	1,596	2,226	1,748
Nebraska	48	118	178	111	116	107
North Dakota	367	422	1,361	1,576	446	661
Ohio	72	707	1,186	325	268	192
Oklahoma	452	312	295	80	125	457
South Dakota	927	131	171	20	53	24
Tennessee	181	95	630	270	81	15
Wisconsin	240	473	197	194	173	273
Gulf Coast (PADD 3)	45,555	38,923	33,652	40,713	35,803	10,729
Alabama	1,369	793	2,014	2,203	2,124	1,649
Arkansas	284	421	283	447	458	329
Louisiana	11,319	940	1,619	3,892	2,900	1,648
Mississippi	722	105	769	845	1,280	1,726
New Mexico	839	548	582	306	859	572
Texas	31,023	36,115	28,385	33,020	28,183	4,805
Rocky Mountain (PADD 4)	950	1,230	1,063	2,267	2,537	663
Colorado	553	761	715	922	1,166	365
Idaho	247	23	40	17	9	0
Montana	145	424	130	109	8	1
Utah	6	23	96	917	1,097	101
Wyoming	0	0	81	303	257	197
West Coast (PADD 5)	105,090	148,447	122,525	122,342	119,723	73,457
Alaska	6,835	7,305	7,875	6,369	7,234	5,748
Arizona	1,237	1,611	1,291	1,649	330	284
California	11,984	10,093	16,107	56,828	36,038	44,068
Hawaii	75,189	1,940	8,939	48,583	2,141	1,557
Nevada	1,868	1,560	1,684	1,745	1,983	1,205
Oregon	2,027	1,948	2,198	2,388	1,469	1,940
Washington	5,950	123,989	84,431	4,781	70,527	18,657

In thousand gallons. Data source: EIA 2013b.

End Notes

[1] The EPA considers co-processing to occur if both petroleum and biomass feedstock are processed in the same unit simultaneously.

[2] EIA. (2013). "Petroleum and Other Liquids - Definitions, Sources and Explanatory Notes," Accessed August 2011: http://www.eia.gov/dnav/pet/TblDefs/pet_cons_821dst_tbldef2.asp.

[3] In July 2011, President Barack Obama proposed that the CAFE standards for cars and light trucks increase to 54.5 miles per gallon (mpg) by 2025 (currently 30.2 mpg for passenger cars and 24.1 mpg for light trucks), the biggest increase since the federal government started regulating fuel economy in the 1970s. The auto industry will be given time to adapt to the proposed 2025 standard. Under the supplemental notice of intent, passenger cars are required to increase fuel economy from 2017-2021 by 4.1% annually while light-duty trucks are required to have 2.9% annual improvements.

[4] EIA. (2013). "Petroleum and Other Liquids - Definitions, Sources and Explanatory Notes," Accessed August 2011: http://www.eia.gov/dnav/pet/TblDefs/pet_cons_821dst_tbldef2.asp.

[5] These residues include unutilized wood volume from cut--or otherwise killed--growing stock from cultural operations, such as pre-commercial thinnings or from timberland clearing. This category does not include volume removed from inventory through reclassification of timber land to productive reserved forest land.

[6] Other forestlands are defined as incapable of producing at least 20 cubic feet per acre per year of industrial wood under natural conditions because of a variety of adverse site conditions including poor soils, lack of rainfall, and high elevation.

[7] Primary mill residues include wood materials (coarse and fine) and bark generated at manufacturing plants (primary wood-using mills) when round wood products are processed into primary wood products like slabs, edgings, trimmings, sawdust, veneer clippings and cores, and pulp screenings (USDA – Forest Service). Secondary mill residues include wood scraps and sawdust from woodworking shops such as furniture factories, wood container and pallet mills, wholesale lumberyards, and flooring.

[8] This category includes wood that has a commercial value for other uses but is used as an energy feedstock because of competitive market conditions.

[9] This volume includes logging residues and other removals, unused primary mill residues, secondary mill residues, and urban wood waste.

[10] NAS (2009) evaluates corn/wheat residues and hay only, while Milbrandt (2005) considers residues from 18 major agricultural crops.

[11] In 2013, biomass-based diesel mandate was revised upwards from 1 billion gallons to 1.28 billion gallons actual volume (EPA 2013b).

[12] West Texas Intermediate crude oil price: $100.54/barrel.

INDEX

A

B

C

D

E

I

J

K

L

M

Q

R

S

T

U

V

W

Y